About Island Press

Since 1984, the nonprofit organization Island Press has been stimulating, shaping, and communicating ideas that are essential for solving environmental problems worldwide. With more than 1,000 titles in print and some 30 new releases each year, we are the nation's leading publisher on environmental issues. We identify innovative thinkers and emerging trends in the environmental field. We work with world-renowned experts and authors to develop cross-disciplinary solutions to environmental challenges.

Island Press designs and executes educational campaigns in conjunction with our authors to communicate their critical messages in print, in person, and online using the latest technologies, innovative programs, and the media. Our goal is to reach targeted audiences—scientists, policymakers, environmental advocates, urban planners, the media, and concerned citizens—with information that can be used to create the framework for long-term ecological health and human well-being.

Island Press gratefully acknowledges major support of our work by The Agua Fund, The Andrew W. Mellon Foundation, The Bobolink Foundation, The Curtis and Edith Munson Foundation, Forrest C. and Frances H. Lattner Foundation, The JPB Foundation, The Kresge Foundation, The Oram Foundation, Inc., The Overbrook Foundation, The S.D. Bechtel, Jr. Foundation, The Summit Charitable Foundation, Inc., and many other generous supporters.

Natural Defense

Natural Defense

Enlisting Bugs and Germs to Protect Our Food and Health

Emily Monosson

Washington | Covelo | London

Library of Congress Control Number: 2016957614

Printed on recycled, acid-free paper

Manufactured in the United States of America
10 9 8 7 6 5 4 3 2 1

Keywords: antibiotics, pesticide resistance, microbiome, genomics, viruses, blight, GMO, pathogen, phage, fecal transplant, machine learning

Contents

Acknowledgments

I am thankful to all of those who supported this project with their encouragement, time, expertise, and editorial skills. I consider this a hopeful book that is based upon the research of scientists, health care workers, and farmers working to improve human, environmental, and agricultural health. To those in particular who took valuable time from their work to share their knowledge and then to read through chapter drafts, I am grateful. Without you there would be no book: Christopher Appleton, Nicholas Bergman, Claude Boyd, Bruce Caldwell, Dan Chellemi, Jon Clements, Joe Crabb, Bill Fry, Brian McSpadden Gardener, Jack Gilbert, Tom Gordon, Cliff Hatch, Brad Higbee, David Hughes, Keith Jones, Randy Kincaid, Steve Lapointe, Margaret Lloyd, Cam Oehlschlager, Kathleen McGraw, Lenny Moise, Justin O'Grady, John Pickett, Margaret (Peg) Riley, Marcel Salathé, Joseph Schwartzman, Seila Selemovic, Peter Reczek, Arthur Tuttle, Ryan Voiland, Jack Vossen, and a few who wish to remain anonymous. Any mistakes in the material or interpretation of the science are my own. A special thanks to Karen Anderson for sharing her and Tim's experience; and Suzanna from Chicago for sharing hers. From concepts to comma, Emily Davis, Island Press editor, has contributed to this project in countless ways, turning clunky phrasing into smooth streams of words while wading through a flurry of e-mails with the subject header "Wait! Read *this* instead." Thank you, Emily. In addition, I am thankful for the assistance of others at Island Press (Sharis Simonian, Katharine Sucher, Jason Leppig, and Jaime Jennings) for their help in moving this book-writing process along. The writing also benefited from the careful copy editing of Michael Fleming, who raised plenty of good questions. While you can't judge a book by its cover, I wouldn't mind if they did in this case, thanks to Roberto de Vicq's cover design.

Throughout the writing process, reviewers willing to point out inconsistencies, too much science, or not enough explanation were essential. Those who spent their precious spare time with nothing to gain, not even a bottle of wine, include Bernadette Albanese and Suzanne Epstein, Sofia Echegaray, Julie Kumble (and her husband Bruce for allowing me to share his experience with antibiotics), Sophie Letcher, Brent Ranalli, Rony Sebok, Bob Strong, and Missy Wick. Scott Russell Sanders's Bread Loaf writing group of 2015 provided encouragement that the topic and material was worth pursuing. I am grateful that the Lady Killigrew Café in Montague, Massachusetts, welcomes those of us who sit and write all morning for the price of a couple of cups of coffee; so thank you, Ella, for keeping it that way. I am thankful for friends like Leigh Rae who kept me fit and ready to sit and write; and for the gift of a nearly lifelong friendship with Penny Shockett, who can not only check my science but my grammar as well; and grateful to Ben Letcher, my partner in life who has listened patiently as I talked and talked (and talked) my way through chapter ideas, quandaries, and second thoughts about material for this book; who provided insightful and careful review of chapters; and who never once conveyed anything but encouragement when, on those long drives, I would turn off the radio and ask him to listen as I read aloud. I am also grateful to my parents for their support in ways that, even now, I am sure I do not fully realize. Thank you, Mom. I would have loved to discuss this book with my father, an early tech guy, a Trekkie, and someone interested in the future. And finally, to Sam and Sophie, who are both now old enough to read and comment on *my* work, love and hope.

Introduction

THIS IS A BOOK ABOUT SOLUTIONS. A couple of years ago, I gave a presentation about the problems of modern agriculture and medicine—specifically, how we are losing our edge against pests and pathogens as they develop resistance to pesticides and antibiotics. Afterward, an audience member asked, "So what can we do?" I shrugged and said, "Use less." There was a little laughter and then an expectant pause, but I had nothing more to say. How could we cure disease without overusing broad-spectrum antibiotics? Or protect crops from bugs and weeds while using less pesticide? This book is how I would have liked to answer that question.

There is little doubt that chemicals have played an enormous role in growing food and protecting against disease over the last century. Pesticides and fertilizers (along with other agricultural practices) have helped farmers ramp up production to feed billions. Here in the United States, we are so certain of abundance that we throw away about 40 percent of all food produced, and we expect beautiful, blemish-free fruits to be available year-round. We have grown similarly accustomed to the miracle of antibiotics. Before the antibiotic age, infections—from meningitis to strep and staph—reigned as all-too-often-incurable killers. Penicillin saved countless lives, and when it failed another antibiotic took its place. Now we demand that our doctors prescribe antibiotics at the first sign of a cough.

Most of us alive today are beneficiaries of this chemical warfare waged by humans against pest and pathogen. It worked, for a while. Then came resistance and other unintended side effects, from altered ecosystems to the emergence of opportunistic diseases: a young man is left to struggle with a drug-resistant infection that has out-competed his normal intestinal flora; a blight responsible for the Irish potato famine now increasingly resists fungicides; aggressive weeds crowd out crops; and common pesticides kill off even the beneficial insects. How do we replace twentieth-century pesticides now fallen from grace, or save our antibiotics so that they are there for us when we most need them?

Fortunately, imaginative strategies are in the works. Some, like gene editing, are of the twenty-first century; others, such as the ancient practice of fecal transplants, are just now emerging from history's shadows, made new again by advances in technology and a sense of urgency. Many strategies are borrowed from nature, one of our best allies against these age-old enemies. There are viruses that infect and bust apart bacteria. Strategic crops create healthy microbial communities able to fend off plant pathogens. There are vaccines engineered to better stimulate our natural defenses, and there are plants engineered to resist diseases with genes borrowed from related strains. Insect pheromones—natural and very specific chemicals—frustrate moths that would otherwise infest our fruits and nuts with their wormy larvae by throwing them into a misguided sexual frenzy. And bacteria provide new kinds of highly selective antimicrobials—targeting pathogens while leaving our microbiomes intact. There are hundreds if not thousands of reasons for cautious optimism; I have selected a handful.

We are also realizing that medical and agricultural solutions go hand in hand because people and plants actually have a lot in common. Whether we are talking about food, the environment, or people, good health is dependent on common biological and ecological factors. A fecal transplant in humans is little different from encouraging soil microbes on the farm. Viruses that infect bacteria are useful both in humans and in the field, as is the concept of using natural enemies for protection against pest and pathogen. And an ounce of prevention is worth well more than a pound of cure, whether we are protecting kids or crops. For the most part, a revolutionary approach in the hospital is a revolutionary approach on the farm. This commonality is why I have paired this book's chapters, exploring each solution's application in health care in one chapter and its application in agriculture in the other. Even when the technologies or treatments aren't identical, I think it is still useful to consider them in tandem. For too long we have thought of ourselves as separate from the environment. The sooner we begin working with, rather than against, nature for our food and health, the better off we will be.

In the process of researching this book I sought out scientists working on cutting-edge solutions, often informed by our increasing appreciation for the intricacies of ecology. From genomics to computational biology to new developments in virology and bacteriology, it is easy to become entranced by research that promises to reduce pesticides and cure disease. So this book comes with a caveat: some of these approaches may well save our food and medicine, but others will fail. One early reader cautioned me that stoking excitement about new technologies too often backfires, shredding the public's trust in science. "It's like writing about the stock market. Suppose you say this is very cool—the latest, most exciting development, the best thing since sliced bread. Then suppose later there is a conspicuous failure; perhaps a good idea doesn't hold up in the field or in the hospital, or doesn't even make it to market." By writing about promising developments, it is not my intention to endorse any single solution as the next best thing but rather to provide examples of scientific ingenuity. While some of these approaches won't pan out, they may inspire a vaccine, or natural pesticide, or probiotic that *does* work. That is the nature of science. It is an enterprise that builds and corrects and moves ahead. We live in an on-demand world and we are impatient for the next miracle cure. But science does not work that way. One new vaccine won't save us. Nor will a new, all-natural pesticide. This book isn't about going all organic, and it certainly isn't about rejecting antibiotic treatments. Instead, it is about moving, step by step, away from our chemical-soaked past and into a future that is more in tune with nature.

All together, these natural defenses—from maintaining microbiomes to enlisting viruses to frustrating insects—have given me a sense of optimism that I want to share with readers. We no longer have to be stymied by the question *What we can do?* Better ways exist to treat pests and pathogens; to reduce our dependence on synthetic chemicals; to stay healthy and grow food. Some solutions are being used today; others won't be available until tomorrow. But hope for a healthier future is here now.

CHAPTER 1

Natural Allies: Our Bacterial Protectors

FIVE YEARS AGO, Tim Stoklosa caught a cold. He was twenty-six at the time, his lungs compromised by muscular dystrophy, a neurodegenerative condition that he has managed for much of his life.[1] Because Tim lacked the ability to cough and clear his lungs, colds predictably led to pneumonia. This time around, it was silent pneumonia. Tim was given Augmentin, a powerful combination of penicillin and clavulanic acid, an enzyme inhibitor aimed at amoxicillin-resistant bacteria. As a so-called broad-spectrum antibiotic, it kills not only harmful bacteria but also plenty of beneficial species that make their home in our gut. For many of us, a few cups of yogurt or some probiotics help rebuild that microscopic community.

But ten days after Tim's first course of Augmentin, his fever continued and he developed stomach upset. He was given another course. It didn't help. "Finally," says his mother, Karen Anderson, a single mom who has devoted a large part of her life to Tim's health, "one nurse figured he had *C. diff.*" *Clostridium difficile* is a potentially lethal infection of the colon. Though the bug may lurk in our guts without causing harm, it is also a notorious opportunist often acquired in the hospital; in the past few years, a particularly dangerous strain has emerged. Wiping out the beneficial gut flora provides the pathogen with the perfect opportunity to set up shop. "When you have *C. diff*," says Karen, "it is like the lining of your colon is coming out of you. It's horrifying." Infection with *C. diff* is a direct consequence of waging chemical warfare against a community of bacteria when our real aim is just to target a few troublemakers. Not only is *Clostridium* an opportunist, but it is particularly difficult to eradicate from the body and from hospital surfaces. Some strains are antibiotic-resistant, and all can form spores—capsules capable of resisting chemical treatment and which can lie in wait for months for just the right conditions to go forth and multiply.

To combat Tim's infection, doctors prescribed a course of Flagyl (metronidazole). It seemed to do the trick. But as soon as the drug left his body, the *C. diff* returned. Next up was vancomycin. In many cases it is the one remaining weapon in the antimicrobial arsenal—a so-called drug of last resort. Tim was on and off "vanco" for over a year. But the resilient bug held its ground. After each ten-day course, *C. diff* returned. "Tim is in a wheelchair, on a ventilator," says Karen, "and he's tolerated the treatments and the ongoing *C. diff*. He's tough." But as the infection dragged on, Tim and Karen became more and more desperate for an actual cure.

Tim isn't alone. The bug causes nearly half a million infections in the United States alone, with nearly 30,000 patients succumbing within a month of diagnosis.[2] Most affected are the elderly and the immunocompromised, but cases are increasingly occurring in the very young who have had no prior antibiotic exposure.[3] In the industrialized world, *C. diff* is the leading cause of hospital-acquired diarrhea and colitis.[4] The pathogen is on the rise—and we are to blame. When we destroy a functioning and diverse ecosystem, we cannot expect it to rebuild itself as it was. Yet that is what we do every time we use broad-spectrum antibiotics.

My own kids were prescribed round after round of bacteria-busting drugs for all sorts of common childhood ills. Even with some background in microbiology, I gave little thought to the havoc caused by those antibiotics; I just wanted my child cured (whether or not the antibiotics, which are not effective against viruses, did the trick). My son Sam slurped his first dose of sticky, sweet, bubblegum-flavored amoxicillin in 1994. He was six months old. On average, kids in the United States will have had twenty courses of antibiotics by the time they reach adulthood. "You must have felt awful, that first time you gave Sam antibiotics," commented an ecologist friend who teaches about evolution and resistance. His own two-year-old had recently become fussy and had begun pulling at his ears, a situation that would have sent me running to the pediatrician for a fix. "You basically annihilated his bacteria." But I didn't feel awful, I felt relieved.

Five years and several ear infections later, I was teaching medical microbiology to nursing students. It was a "get to know the enemy" kind of class. We examined one pathogen after another: its life history, how and where it attacked, how quickly it reproduced, its favorite conditions, and how a patient might respond to it—the signs of infection. One day the students had an opportunity to become acquainted with some of their own bacteria. They swabbed their skin, mouth, or whatever body bit they dared, inoculated petri dishes, and popped the dishes into an incubator set at body temperature. Days later, like a container of sour cream forgotten in the back of the refrigerator, the plates were covered with growth. Trillions of bacteria piled up in colonies initiated from single cells. There were glistening white dots; globular egg-yolk-like mounds; salmon-colored bubbles. Tiny studded colonies grew next to those with undulating edges or wrinkles. Others oozed across the plates like phlegm. No two plates were alike, each displaying an incredible diversity of microscopic life rendered visible by sheer numbers. Most of the bacteria on those plates wouldn't bother us, and many are beneficial. But some, given the opportunity, would make us sick. Even as the students marveled at the number of species growing on and within their bodies, no one thought to ask what happens when we poison the whole lot of the bacteria in an attempt to eradicate a few.

The bacteria on those plates represented a small fraction of the microbiota (bacteria, viruses, fungi, and other organisms) cohabiting the students. If we could take a teaspoon of liquid from our stomach—an organ once thought to be sterile—it would contain several thousand bacteria; if we were to do the same in our colon it would be more like 100 billion.[5] Our gut—from intake to exit ramp—has a surface area equal to the floor space of an efficiency apartment, and it is coated with bacteria.[6] Skin has the surface area of a playground's four-square court, though it too plays host to a universe of microscopic life. Our bodies support thousands of *different* bacterial species, and we carry around more bacterial cells than human cells. Most of those bacteria have shared their lives with humans for eons. Some pass from mother to child. The first time I fed Sam that amoxicillin, I didn't know that I would also be killing off bacteria that likely descended from my own microscopic flora. When he was born, he picked up my bacteria as he made his way from uterus to vagina and into the world, and he ingested still more as he latched on to my breast, hungry for mother's milk. Much of this natural microbial ecosystem was disrupted months later with a single teaspoon of pink liquid.

We pay little attention to these casualties, the harmless and beneficial bacteria lost alongside the pathogens. Plenty of us pull through just fine, eventually recolonized by survivors or other sources of bacteria. Some of us may benefit from probiotics, mixtures of living bacteria that are able to reseed the bacterial turf. And then there are others, like Tim, for whom the cost of disrupting this microbial ecosystem can be life-threatening. Roughly seventy years ago, we began using antibiotics on an industrial scale. Yet we are only now realizing the destruction we have wrought upon our own vital ecosystems. How can that be?

MicroGenetics

Within and upon our bodies, microbes outnumber our own cells. Just in terms of bacteria alone, according to a cell-for-cell recent estimate, our bodies are roughly one part human and one part bacterium.[7] These microbial species intermingle, form biofilms, spit out toxins, have sex, and produce clone after clone. Our lives depend upon an invisible diversity of bacteria, fungi, protozoa, and viruses, but we know little of these microscopic allies. We knew practically nothing of them until the seventeenth century, when Dutch draper and scientist Antonie van Leeuwenhoek discovered "animalcules" under his microscopes. Fashioned from lenses originally intended to help distinguish good textiles from bad, Leeuwenhoek's microscopes introduced humanity to the bustling world of microbial life. It would be another 200 years until German physician Robert Koch developed his elegant "postulates" of disease causation, providing criteria to link a disease to a particular microbe, before medical microbiology finally hit its stride. Within a few years, diseases like anthrax and tuberculosis could be pinned on specific bacteria. Applying Koch's postulates required isolating and growing pure bacterial colonies. A drop of blood, saliva, or even soil was smeared across a nutrient-rich agar, and within a day or so pinprick colonies began to bloom as individual bacteria divided again and again until visible colonies emerged. Culturing and identifying bacteria is an art that has changed little since Koch's day. Walk into a microbial laboratory today and you are hit with the smell of agar and nutrient broth like a blast of concentrated chicken soup.

Until recently, our knowledge of microbes remained limited to those species we could capture and culture: an estimated 2 percent of bacterial life. But from that fraction we learned that bacteria have circular chromosomes (unlike plants, animals, and humans, whose chromosomes are linear), that fungal cells are much like our own, that chemicals excreted by bacteria and fungi can be used as antibiotics, and that bacteria can trade genetic information directly from one microbe to another. We also learned that all life—whether bacterium, barnacle, bed bug, or human being—shares a genetic code. Just four different molecules—guanine, cytosine, thymine, and adenine, denoted by G, C, T, and A—spell out the proteins that build our cells and provide directions for their production, in large part making us what we are.

Since the early decades of the twentieth century, scientists have understood that DNA was somehow responsible for carrying traits from one generation to the next. This realization led to a whirlwind of discovery. By mid-century, Rosalind Franklin, George Watson, and Francis Crick revealed the very structure of DNA: the now-familiar winding ladder of paired bases known as the double helix. A few years later, in 1957, Crick proposed that the genes encoded by DNA directed the synthesis of proteins by controlling the order in which amino acids link together. Within the next two decades, scientists were learning how to cut DNA apart and paste it together—the first steps toward genetic engineering. Yet, as exciting as the advances were, *sequencing* DNA, or determining the precise order of those bases, remained a tedious process. Scientists could read enough code to sequence the amino acid building blocks of proteins, but decoding a simple virus remained beyond reach. Scientists were limited to browsing the children's section of the genome library while dreaming of Tolstoy and Proust. But now, new technology is providing access to almost every book in life's library, augmenting what we know about everything from genetics to evolution, ecology, and microbiology. And this knowledge is fueling a revolution that promises new solutions to age-old problems, both in the hospital and on the farm.

The key to genetic speed-reading began with advances by teams led by Walter Gilbert and graduate student Allan Maxam at Harvard University, and Fred Sanger at Cambridge University in England. Working nearly simultaneously, both groups made breakthrough discoveries that, relatively rapidly, allowed scientists to bring order to life's A's, T's, G's, and C's. (The work garnered Gilbert and Sanger a Nobel in 1980; it was Sanger's second.) Though their approaches differed, their efforts kick-started the present era of genetic sequencing. Genes, which may run as long as 10,000 bases, could finally be deciphered in a reasonable amount of time, enabling scientists to read the paragraphs of life. But still, the full story remained beyond their reach. Pages were out of order. And much of the "literature" seemed garbled: a tangle of genetic gibberish that scientists labeled "junk DNA." Progress was still relatively slow. But meanwhile, another technology that would change the world was also developing rapidly: computer science.

In the 1970s, when Gilbert and Sanger were working out their techniques, each bit of information had to be transferred from laboratory notebook to computer punch cards and then typed up for publication while scientists spent days and nights waiting for results.[8] It was tedious and expensive. Even so, DNA sequences were proliferating at such a fast clip that the National Institutes of Health (cosponsored by various health institutes including the National Cancer Institute, along with the Departments of Energy and Defense) created GenBank: the world's first computerized database for the code of life. Over the course of just a few years, as ease of sequencing combined with increased access to computing power, and as punch cards gave way to magnetic tapes, the size of the database grew more than twenty-fold. By 1985, GenBank held around 5,700 sequences—decoded stretches of DNA describing bits of viruses, plants, and animals that constitute some of the first genetic representations of life on earth.[9] At that time there were 570 bacterial sequences, or a total of nearly 700,000 base-pairs (pairs of the letters G, C, T, and A). These were bits of sequenced DNA from bacteria, but not *wholly* sequenced bacteria. Despite all the decoding and reporting, gaps remained. As the twentieth century was coming to a close, no living organism had been fully sequenced.

Hemophilus influenza—a pneumonia-causing microbe possibly similar to the bug that forced Tim onto antibiotics—would be the first organism fully sequenced. Craig Venter and others cracked its genetic code in 1995, translating the 2,000 protein-encoding genes and nearly 2 million base-pairs. It would be the first "fully translated" book in life's genomic library. But five years earlier, an even more audacious project had begun: the Human Genome Project, one of the largest global biological collaboratives at the time. Initiated by the US Department of Energy, the National Institutes of Health, and international collaborators, the project had been creeping along. The slow pace, combined with differences in opinion about methodologies, frustrated Venter, who was fast becoming a sequencing revolutionary. Pushing the pace of discovery, Venter founded Celera, a privately funded company. In 2001, the Human Genome Project and Celera published simultaneously in separate journals a first "draft" of the human genome. Humans had finally cracked their own code.

Sequencing the human genome pushed genetic sequencing technology from a sluggish, labor-intensive, and expensive prospect to an automated process where, for pennies, tens of thousands of DNA base-pairs can be sequenced in a few hours or even minutes. "What used to take thousands of culture plates, now takes one tube. For five bucks, 100 different populations are revealed," says one microbial ecologist who for decades has isolated and cultured soil microbes one population at a time. Much of the publicly available genetic data is deposited into GenBank, which now holds information for hundreds of thousands of species. It is the place to go if you're looking for a sequence or have one to report. The vast majority belong to microbes: bacteria, viruses, archaea (bacteria-like organisms). Sequences for tens of thousands of bacterial species (some genomes more complete than others) are now available.[10]

For decades we have known that we are more than just an organized collection of animal cells. But new sequencing technologies are empowering microbiologists to seek out life beyond the limits of the agar, revealing complex microbiomes in humans, soils, the deep sea, extreme environments, and elsewhere. The library is open, with billions of books to be read. But there are stark differences between reading and understanding. We may be able to speed-read genomes, but the attributes of each newly discovered virus or fungi or bacterium remain to be teased out of masses of data. What fuels their growth? With whom do they associate? Under what conditions might they become pathogenic—or beneficial? Metagenomics, the sequencing of *all* the genomic DNA in a community, can help.

By providing unprecedented insights into microbial communities, metagenomics is changing how we think about life on earth and elsewhere. Focusing on the single microbe is no longer sufficient. Just as humans live in neighborhoods that are defined neither by the delinquent who steals change from unlocked cars nor by the generosity of the neighbor who shares her homegrown tomatoes, but rather by the rich diversity therein, so too microbes exist in complex communities. The more we know about who does what, the better. Metagenomics not only opens the door but may someday provide insights into what makes a community hum, and what may rip it apart. It is the next step in the genetic technology revolution, and it is already shaking things up in a big way. Metagenomic technology "gives us the ability to explore the microbial world in much higher resolution," says Jack Gilbert, a microbial ecologist at the University of Chicago and founder of the Earth Microbiome Project. Gilbert explains, "It's like looking at the stock market. If we only had information about the market at the end of each day we'd have a low-resolution understanding of how things change. It's all about fluid dynamics in space, time, metabolism, and functional composition."[11] Metagenomics is enabling us to capture ecology in motion, both in our bodies and in the world at large. That is a powerful thing because we are now realizing that, despite all their benefits, our antibiotics and antimicrobials have their downside. Wholesale destruction of bacteria, whether in the human body or the agricultural field, can be profoundly disruptive.

Chemical and Biological Weapons

Humans have been at war with some kinds of bacteria and viruses since the beginning of our existence. Waves of the Black Death swept across Eurasia, anthrax infected livestock, and the blight that infected potatoes contributing to Ireland's Great Famine. Yet even before we knew much of anything about the genetic code or the microbial world, we knew how to protect ourselves. Well before viruses were known to exist, for example, smallpox was held at bay with crude vaccines. Young women in the maternity ward were saved from postpartum death when physicians started washing their hands in between patients, though at first the doctors had no knowledge of bacteria like *Staphylococcus*, *Streptococcus*, and *Clostridium* that might have been hitching a ride on their skin. And, over a decade before Koch developed his postulates, efforts to keep human waste away from sources of drinking water held cholera in check. The revelation that invisible microbes could not only infect a body but take a life must have been humbling and disturbing. But Louis Pasteur saw hope in these discoveries, writing in 1878: "If it is a terrifying thought that life is at the mercy of the multiplication of these minute bodies, it is a consoling hope that Science will not always remain powerless before such enemies."[12] Within a few decades, his prediction proved true.

Antimicrobials (which include antibiotics) mark a milestone in the ongoing conflict between humans and pathogens. But, because the battleground is our body, chemical attacks against microbes run the risk of collateral damage to our own cells. Early antimicrobials such as mercury and arsenic were notoriously toxic. So how to destroy the pathogen without killing the patient?[13] This biochemical quandary was solved incidentally by microbiologists simply seeking a better view.

Experimenting with the vivid blues, reds, and purples produced by the booming nineteenth-century synthetic dye industry, Hans Christian Gram sought to distinguish the pathogens infecting diseased lungs from human lung cells. Viewing live cells with a *modern* compound microscope is hard enough; distinguishing one from another is nearly impossible. But loading a cell with dye can make all the difference. Using a combination of available dyes, Gram managed to infuse killed bacterial cells with color. Some stained blue, while human cells remained *au naturel.* This selective staining captured the imagination of German physician Paul Ehrlich, who wondered: If chemicals could distinguish microbe from human, could they do more than that?

Capitalizing on the selective power of those turn-of-century dyes, Ehrlich sought out molecules that killed microbes while leaving our own animal cells unharmed. After testing hundreds of candidates, Ehrlich and colleagues hit the jackpot: a chemical that zeroed in on an obstinate scourge, syphilis. Caused by *Treponema pallidum*, syphilis killed and maimed men and women, as well as children born to infected mothers. Marketed in 1910 as Salvarsan, Ehrlich's synthetic antimicrobial was a finicky breakthrough drug. Most of the time it cured, but was difficult to administer safely. Sometimes the treatment was lethal.

This was the case with Margaret K., an eighteen-month-old whose fair-haired image haunts the pages of the *Journal of the American Medical Association*. Her vulva and anus were covered with syphilitic warts and tumorous growths, lesions swarming with *Treponema* spirochetes likely passed along to her from her mother. Salvarsan worked like magic. "It is almost impossible," wrote her physician, "to credit sufficiently the immediate benefit . . . by those who saw her day to day."[14] Then toxicity set in. The drug seemed to drain her of strength: her heartbeat became faint, her legs became flaccid, she grew too weak to lift her head. Her physician lost hope, writing, "The infant showed all the evidence of toxicity. . . . The prognosis is dubious."[15]

Still, when the drug worked, the effects were nearly miraculous. The pathogen that had driven men and women mad, if not to their graves, had been conquered. Ehrlich's discovery, though limited to syphilis (today, such selectivity would be viewed as a *good* thing), opened the chemical front in the war against infectious diseases. How many more pathogens would fall, as humanity's new command of atoms and molecules flourished? Sulfa drugs came next. Discovered in the early 1930s by German scientist Gerhard Domagk, these new medications bested Salvarsan by killing a broad range of disease-causing bacteria. For the first time in human history, a number of once-fatal infections could be cured with a single chemical.[16] Nowadays, it is difficult to imagine the impact these discoveries must have had on a populace accustomed to death from a puncture wound, or smallpox, or a respiratory infection. And it is only with today's hindsight that we recognize the flaw in this broad-spectrum strategy.

Sulfa drugs ruled for nearly a decade, at times losing ground to evolved resistance, but then a new, even more effective drug came along: penicillin. Its discovery by Scottish scientist Alexander Fleming is the quintessential story of serendipity: a scientist gone away on vacation, petri dishes left to molder in the sink, and a subsequent observation. But more than a decade would pass before penicillin hit the pharmacy shelves. With improved technology, combined with 1940s wartime desperation and the discovery of a far more productive *Penicillium* mold (found growing on a cantaloupe in Peoria, Illinois), Fleming's find eventually turned into one of the greatest discoveries in modern medicine. Unlike Domagk's or Ehrlich's discoveries, this was a potent broad-spectrum killer, produced by nature and harvested by humans. No longer would humanity be the helpless victim of pathogenic bacteria.

After penicillin came other antibiotics that are, by definition, antimicrobials produced by living things: aminoglycosides, carbapenems, cephalosporins, macrolides, and others. Yet as much as our concerns about ingesting chemicals might be assuaged by defining antibiotics as *nature's* chemical weaponry (*natural* must mean safe, right?), the reality may be quite different. In nature, bacteria, fungi, viruses, and single-celled protozoa coexist in complex communities. Conflict over territory and resources is natural, but so too is communication, cooperation, and organization. While some of the chemicals we call antibiotics may be used to defend territory, or food, others may serve as means of communication. In our quest to eradicate pathogens, we have paid little attention to how these societies of microbes might regulate themselves. Our practice of culturing individual bacterial species has obscured the complex nature of microbial communities. This is like pulling aside the so-called problem child in a classroom, without considering how her schoolmates and teacher might have influenced her behavior. For nearly a century, we've employed broad-spectrum antibiotics to great effect, saving countless lives. They are truly miracle cures. But we have also used them with a limited understanding of their consequences, particularly to those who may undergo repeated courses of treatment, or to those afflicted with hospital-acquired infections. As a result, Tim, and tens of thousands like him suffer from post-antibiotic infections like *C. diff*—opportunistic bacteria that thrive like looters after a disaster as a once-orderly community is thrown into disarray. How do we right this imbalance?

Missing Microbes

In the early days of microbe discovery, as one pathogen after another was revealed, even Louis Pasteur knew there was some good in the microbial world. "Life would not long remain possible," wrote Pasteur, "in the absence of microbes."[17] Could it? Taking up the germ-free challenge, scientists eventually developed "gnotobiotic" animals (that is, animals living free of microbes). Rats, guinea pigs, and chickens have all survived germ-free. Mice living the germ-free life not only survived, but lived longer. Considering the human lives lost to pathogens, the idea of life without microbes, and without disease-causing microbes in particular, was tempting.[18] Medical microbiology began with a focus on the pathogens: the only good bacterium was a dead bacterium. We have lived by that edict for a century. From our antibiotics and antiseptics to the recent spate of antimicrobial-infused products and the constant availability of hand sanitizers, we have become germophobes. This concept of better living without microbes has left us ignorant of what many now refer to as a vital organ: the human microbiome. That we harbor loads of bacteria within and upon us has been known for decades, but what all those bacteria were doing in our guts and on our skin, no one really knew. We focused on the *micro* while neglecting the *biome*.

Martin Blaser, physician and director of the Human Microbiome Project at NYU, warns that our lifestyle is causing us to lose bacterial species that have been with us for ages. In his book *Missing Microbes*, Blaser writes about the differences between infants born vaginally and those of infants born through cesarean section. While the microbial makeup of infants who make their way through the birthing canal tends to resemble that of the vagina, the latter look more like the assemblages of bacteria residing on mom's skin. Although exposure to the outside environment eventually seems to bring their systems in line, Blaser wonders about possible lasting effects caused by these early-life differences. And then there is the case of the disappearing *Helicobacter pylori*. For years, physicians in training as well as scientists were taught that our acidic stomachs were sterile. But *H. pylori* lives in the guts of about half the human population. A two-faced bacterium charged with causing stomach cancer and ulcers but also with maintaining our immune system, *H. pylori* treads a middle ground between friend and foe.[19] Blaser writes that virtually all children used to carry it, at least early in life. But today, as children—including my own—are treated with one course of antibiotics after another, *H. pylori* is disappearing. "In its protected gastric niche for eons," writes Blaser, "*H. pylori* was not at all prepared for the onslaught of antibiotics in the last seventy years."[20]

It is undeniable that antibiotics have been and continue to be lifesavers. But at the same time, by poisoning our own bacteria (and by living in an increasingly hygienic world), we have altered the human microbiome. And once an ecosystem or community has been altered, restoration is a tricky business. We see this in forests, coastal regions, wetlands, and now in our own guts. In Tim's case, while the Augmentin busted apart the bacteria infecting his lungs, the drug likely killed off myriad other bacteria including lactobacilli, *E. coli*, *Proteus*, *Klebsiella*, *Enterococcus*, *Bacteroides*, and *Helicobacter*.[21] All are members of a complex community that may prevent *Clostridium difficile* from settling in.

As a class of microbes, *Clostridium* are a nasty bunch. Some of us of a certain age might be familiar with *Clostridium botulinum*; once common in improperly canned food, it is the reason we check the pantry for bulging cans. A gram of botulinum toxin, distributed about, can kill millions. *Clostridium tetani* lurks in soils. Lodged deep into a hand or foot via a puncture wound, it can kill. Some strains of *Clostridium difficile*—*C. diff*—produce toxin, too. They can also form spores, allowing them to weather unfavorable conditions due to lack of food, or temperatures that would typically sterilize, or the presence of antimicrobials. Clostridia are obstinate. Expelled during a bout of diarrhea, *C. diff* is easily transmitted. Moreover, *C. diff* spores can linger on countertops, linens, toilet bowls, or door handles, which is why, in part, it has become a scourge of hospital patients like Tim.

First described in 1935 as part of the "normal intestinal flora" in infants, *C. diff* rose to prominence in the 1970s when physicians recognized its role in antibiotic-associated diarrhea.[22] Toxin-producing strains can wreak havoc on cell structure and function, causing bloody diarrhea, intense pain, and, in severe cases, extensive destruction that requires surgical removal of damaged bowels.[23] As Karen explained, it may seem as if your whole colon is coming out. In 2011, nearly half a million Americans suffered a *C. diff* infection, at least a quarter of these acquired while in the hospital.[24] Twenty-nine thousand died within thirty days of diagnosis.[25] While Tim may or may not have harbored a resident population of *C. diff*, it is likely that his repeated visits to the hospital didn't help. And, after a year of treatment, it became clear that vancomycin was not the solution.

Repopulating

More than a century after Pasteur and Koch, the role of our microbiota has made medical headlines as one study after another reveals that our bodies and our minds are in cahoots with hordes of microbes. Remember those germ-free mice and rats? While their bodies may have appeared healthy, recent research indicates that their organs, including their brains, may have been altered compared with those of normal mice. A number of studies now suggest that gut microbes send messages to the brain.[26] From movement to memory, the lack of a healthy microbiome may influence behavior—not only in mice but in humans as well.

After decades of ignoring beneficial bacteria, we are now becoming so obsessed with them that my husband wonders how long before someone claims the microbiome defense: "My gut was out of whack, so I couldn't stop myself." Microbes are the cure *de jour*—with varying degrees of success—for everything from diarrhea to obesity to mental health. There is talk of probiotics for the gut, for vaginal health, and for skin. But at present, probiotics are the Wild West of health care. As all-natural dietary supplements rather than drugs, they are subject to little or no regulation, including testing for efficacy.

A few years ago, I finally tossed the bottle of *Bifidobacterium* probiotics, which, along with other nutritional supplements like fish oil and assorted vitamins, had colonized the upper shelf of our refrigerator. *Bifido* is a common gut microbe. The bottle was given to me by a neighborhood nurse as a preventative for Sam and his sister Sophie, who, as preschoolers, seemed to be an antibiotics tag-team. Skeptical of treatments that seemed imprecise or unproven, I am now somewhat embarrassed to admit that the bottle was never opened. Fortunately, the kids seemed to weather the amoxicillin, Keflex, and even doxycycline ("doxy") without obvious problems. But who knows what changes were wrought upon their gut flora? In his TED talk on gut microbiomes, Rob Knight, cofounder of the American Gut Project, showed an image of gut flora from a child treated with antibiotics. It was almost like hitting a "reset" button. The flora community, which had been trending from one characteristic of an infant toward that of an adult, crashed. Knight describes the event as "a setback of many months."[27] Watching the talk, I wondered how my children's guts had responded during those "antibiotic years," and whether that unopened bottle of *Bifido* would have made any difference.

Today there is an explosion of probiotics, but not enough conclusive data. Despite all the claims we see in magazine ads and news articles, or read on bottles lining grocery shelves, the National Institutes of Health's Center for Complementary and Integrative Health is circumspect. In 2015, they issued the following statement:

> We still don't know which probiotics are helpful and which are not. We also don't know how much of the probiotic people would have to take or who would most likely benefit from taking probiotics. Even for the conditions that have been studied the most, researchers are still working toward finding the answers to these questions. . . . The US Food and Drug Administration (FDA) has not approved any probiotics for preventing or treating any health problem. Some experts have cautioned that the rapid growth in marketing and use of probiotics may have outpaced scientific research for many of their proposed uses and benefits.[28]

In other words, the science is just too new for anyone to be able to pin down a good microbial brew. But there is one treatment that is gaining traction within the FDA and elsewhere: fecal transplants.[29] Fecal transplants are like untamed probiotics. In its simplest form, a fecal transplant is just that—the transfer of feces from one gut to another. And it's nothing new. Seventeen hundred years ago, during China's Dong Jin dynasty (317–420 AD), health practitioners wrote of administering a suspension of human feces "by mouth" to those suffering from severe diarrhea. The treatment was said to bring patients "back from the brink of death."[30] Today, with fecal-transplant cure rates for *C. diff* ranging from 80 to 100 percent, depending on the study, so-called poop banks are popping up around the country.[31] In the Boston area, you can earn $40 by donating your stool to OpenBiome, a nonprofit devoted to "expanding safe access to fecal microbiota transplants" and "catalyzing research into the human microbiome."[32] After a thorough screening (a clinical interview followed by blood and stool tests to ensure that you aren't carrying any infectious diseases), you are free to start donating.

In 2010, when Tim had become infected, transplants were not standard treatment. But nothing seemed to be helping. "We were willing to do anything," recalls Karen. "We did some research. There were stories from a gastroenterologist who told of people who'd lost fifty pounds [due to infection], who were cured almost instantaneously." At the very least it was worth a shot. Because the treatment wasn't FDA approved at the time, the procedure had to be done under the hospital radar. "We couldn't tell anyone what we were doing," says Karen. "The hospital thought it could be dangerous, they didn't want it performed." But, she says, their doctor was an outside–the-box kind of thinker. A renegade.

Next, they needed a donor. A close relative—a parent, sibling, or child—is best, and Karen was the obvious choice. Unfortunately, family members may be hosting their own pathogens—if not *C. diff*, then others like HIV, hepatitis, or common ailments of the digestive tract, from pathogenic *E. coli* to Salmonella. "We did a lot of testing," says Karen. Then came the training. "You have to perform on the spot." The transplant was at 11:00 a.m., [so] I had to produce at 10:50." Karen's donation was strained, processed, and administered through an endoscope—the same instrument many of us have experienced by way of colonoscopy. There were no guarantees. And even if it did work, Tim would likely be treated again and again with antibiotics. Could a newly implanted community hold up to such abuse?

Within twenty-four hours Tim's severe diarrhea disappeared, replaced by perfectly formed stools. Invoking a sentiment associated with antibiotics nearly a century earlier, Karen says the outcome was miraculous. And, in the five years since, Tim's new flora has managed to fend off *C. diff* even through repeated bouts with infection and antibiotic treatment. Tim is only one of thousands. Increasingly studies are confirming what ancient Chinese practitioners suspected nearly two millennia ago: fecal transplants work. Notably, the treatment can be more effective than vancomycin—by a long shot. Yet even as antibiotic-induced *C. diff* infections are on the rise, the curative power of antibiotics cannot be shelved. "My father's mother died of pneumonia because they didn't have antibiotics," reflects Karen. "People died of pneumonia all the time."

A few years later, in 2013, the FDA became involved, stating that the agency would treat fecal transplants as an experimental drug requiring doctors to submit research applications. Essentially, the agency threw cold water on the practice. But thanks to an onslaught of protest, highlighting the dire need for transplants, the agency backed off within weeks, allowing treatment pending informed consent, need, and full disclosure of risks.

Teach War No More

Even as this elegant solution proves to be lifesaving, we still fail to recognize it for what it is: an act of ecology. After the publication of a study confirming the curative power of fecal transplant versus vancomycin, one blog devoted to evolution posted an article with the headline: "Fecal

Transplants Shown Effective—No Mention of Ecology or Evolution." If ever there was a time to mention the importance of either, this would be it.[33] Fecal transplants are the ultimate ecological solution. Over the past few centuries, we have sought cures to beat nature, but that is an impossible task. Nature *is* us and we are nature.

As we come to grips with the microbial world within us and around us, we are beginning to recognize that we have far more allies than enemies. "Our most sophisticated leap," wrote Nobel winner Joshua Lederberg, observing that our collective genomes are "yoked" together, "would be to drop the Manichaean [dualistic] view of microbes—'We good: they evil.'"[34] Instead, we must accept that we are interdependent. Twentieth-century technology isolated us from nature, but now twenty-first-century technology is repairing the rift. Advances in genomics, computer sciences, analytical chemistry, and other new technologies will lead to better health. Using new technologies informed by ecology, we can work *with* the natural world rather than *against* it. And when there are no other options except to attack, at the very least we can manage microbes more strategically, targeting the harmful ones while leaving the larger community intact. Had Tim's initial infection been treated with antimicrobials able to zero in on pneumonia-causing bacteria, perhaps he would never have spent a year on vancomycin, fending off *C. diff.* We can do better. That is what this book is about.

CHAPTER 2

Natural Allies: Microbial Farmhands

It is late afternoon, June. I squat in the field, picking plump, warm strawberries as the sun beats down on my back. The air is fragrant with berries squashed underfoot. Their juice stains my fingers. I fill the one-quart allotment from my CSA (Community Supported Agriculture), Red Fire Farm, and then stuff my mouth with berries, fresh and candy-sweet. These are beauties, grown without synthetic chemicals, nestled in straw. How much effort, I wonder, has gone into maintaining this half-acre or so of organic perfection? For a New Englander, the berries are harbingers of summer, the first local fruit of the growing season. But the season is easy to miss. It lasts only a few precious weeks and then you are once again dependent on fruit from Florida or, more likely, California: well-traveled berries that must travel well—not nearly as sweet, but good enough. Ever since a postwar boom made California the berry-growing capital of the world, the state has been sending out more berries than any other.[1]

As a soft fruit, berries are delicate compared with apples, and thin-skinned compared with oranges. While we might peel away or scrub pesticide-tinged skins of these larger fruits, reducing residues though not removing them entirely, this isn't an option for berries. That's why, in part, conventionally grown berries may contain a laundry list of pesticides, from acetamiprid to captan and pyraclostrobin.[2] Their chemical load has made berries notorious members of the Environmental Working Group's "Dirty Dozen"—the annual ranking of fruits and vegetables based on detectable pesticides.[3] It's understandable, says plant pathologist Margaret Lloyd, since strawberries are one of the few crops that spend over thirteen months in the ground.[4] That is thirteen months of exposure to wilts, rots, insects, mites, and weeds, all vying for precious water and nutrients. In short, strawberries are a high-maintenance fruit. But someday soon, California's berry growers may turn away from some of their synthetic go-to chemicals in favor of a more natural method of protection—soil microbes. Just as nurturing our own microbiome may reduce the need for antibiotics, culturing healthy soils can reduce plant disease and, therefore, the use of chemicals to treat disease. Beneath our feet, a world of microbes, clinging to soil particles and plant roots, is alive and thriving. This plant microbiome may ease us away from pesticides.

A Microbial Zoo

An article about the plant microbiome in the *Atlantic Monthly* features an arresting image of a pine seedling, sprouting up against a glass window.[5] As with an ant farm, you can see both above- and belowground. The young plant is delicate, the kind you see popping up on the forest floor, looking like a Dr. Seuss creation. Its thin green stalk supports a head of just a couple dozen needles. Belowground is a different story. Though the roots only reach down several inches, an expanse of delicate white branches resembling a big old maple tree extends into the soil. These threads, called fungal hyphae, are tied to vegetation that, like nursemaids, help feed and water the seedling. But there is far more to this image than the eye can see. In and around the soil lives a "rhizosphere zoo" chock-full of bacteria, nematodes, fungi, viruses, protozoa, and more.[6] The numbers are astounding: a billion bacteria, 100 million archaea, a billion viruses, 100,000 fungi and fungi-like organisms, along with hundreds of thousands of other single-celled organisms and a handful of tiny nematodes—all in one gram of soil.[7] There are no barriers or bars in this zoo. It is a wild place rife with predation, cooperation, and warring populations that compete for food and space. Raging within the soil are some of the Earth's most ancient interspecies conflicts. Yet through the process of evolution, these communities have achieved some sort of dynamic equilibrium. And we can take advantage of that.

A plant's microbiome, whether that of a pine tree or a pea, is among the most complex ecosystems on the planet.[8] While the microbial mix on our skin may differ vastly from that in our guts, there is some consistency among us. All our microbes exist in a tightly controlled environment where body temperature, moisture, salinity, and nutrients are relatively constant. Agricultural soils, in contrast, are an open system. There are droughts, heat, cold, floods, and, depending on the crop, a dousing with Roundup and other pesticides and fertilizers. Microbial populations bloom as they scuffle over nutrients or a bit of root tip ripe for colonization, and they fade as they succumb to pesticides or an aggressive neighbor. And although corn, potato, squash, strawberry, or grass seedlings don't acquire a starter microbiome as we humans do, once a plant begins sending roots into the soil something miraculous happens. They begin *recruiting* microbes. Plant roots release sugars, proteins, amino acids, hormones, and other chemicals that attract fungi, bacteria, and other microbes—creating a collaboration between plant and life belowground.[9] Many are harmless, some make essential nutrients more available, and others suck the life from a plant. Then there are the protectors: microbes that suppress disease through direct attack on pathogens or provoke a plant's own immune system, kicking it into action.

Farmers have always recognized a few beneficial microbes. In particular, they rely on nitrifying bacteria to pull nitrogen from the air, transforming an essential nutrient into a form that is available to plants and, subsequently, to animals. But now, genomic technology is expanding the known world. We understand that our own microbiome is but a small fraction of the Earth's collective microbiome: an entity so complex that geoscientist and environmental engineer Thomas Curtis likens it to multiple universes. He writes, "I make no apologies for putting microorganisms on a pedestal above all other living things. For if the last blue whale choked to death on the last panda, it would be disastrous but not the end of the world. But if we accidentally poisoned the last two species of ammonia-oxidizers [nitrifying bacteria], that would be another matter."[10]

Plant pathologists, soil microbiologists, and agronomists are increasingly tuning in to the *rhizobiome*—the vegetal equivalent of our gut microbiome. This collection of microbes inhabiting the root zone is alive and teeming. There is living gold in the ground, and, if it can be cultured, robust microbial communities can benefit growers, workers, local communities, and consumers by reducing the chemical dousing now required to protect a crop. Our agricultural future—our food, including strawberries—may well depend upon cultivating a more positive relationship with this unseen yet thriving world of microbes. So too may the California strawberry industry.

Wilt

Strawberries are a labor-intensive, high-value crop, grown on high-value land. California devotes over 40,000 prime acres to the fruit, while the second-biggest growing state, Florida, plants about 10,000 acres, and Oregon trails with a couple thousand. In 2015, over two and a half *billion* pounds of California berries were picked, packaged, and sold. Worth many billions of dollars, berries are a fruit well worth protecting.[11] But they attract plenty of pests. One common problem is verticillium wilt, a disease caused by *Verticillium dahliae*.

A fungus that transforms vigorous green leaves into dry brown litter, verticillium is a grower's nightmare—or a perverse fairy tale.[12] The pathogen can survive dormant in soil for up to fourteen years, like a sleeping princess waiting to be awakened by a kiss. In this case, secretions from plant roots will do the trick. Once triggered, the fungus germinates and moves into plant roots and eventually into vessels carrying water and nutrients to the shoots. Infected plants essentially choke on the fungus, cutting off their own water circulation in an attempt to isolate it. Unable to take up and distribute fluids, the plants require more water and become susceptible to dry spells. Eventually, they wilt and die. In addition to strawberries, verticillium infects hundreds of other species, from food crops to ornamentals and even trees.

Over half a century ago, as the strawberry industry began to take off, there emerged a new way of controlling pests and pathogens, including wilt: soil fumigation. The fumigant of choice was methyl bromide. Unfortunately, much like a broad-spectrum antibiotic, the fumigant didn't distinguish between good and evil. Beneficial microbes and insects perished as the chemical permeated and effectively sterilized the soil. Used by the thousands of tons every year, methyl bromide became the industry standard and helped build an empire. It was soon combined with chloropicrin, more commonly known as tear gas, and together, the toxic duo killed off nematodes, weeds, and *Verticillium* species. Production improved.

"Methyl bromide is effective," says Margaret Lloyd, "because used as fumigant in gas phase, it moves through soil easily, and . . . strawberries have shallow roots, so you don't need to go deep into soil." However—the chemical is acutely toxic to humans, and while innocent of contaminating our strawberry shortcake (it is applied to soil well before planting, so no residue is detectable in fruit), the pesticide is also guilty of ozone destruction. Though methyl bromide is injected more than a foot deep into the ground, which is subsequently covered with plastic, an estimated 50–95 percent of the fumigant eventually ends up in the atmosphere.[13] Methyl bromide has been on the EPA's hit list ever since passage of the Montreal Protocol, the international treaty intended to protect Earth's ozone layer. But while the treaty called for phase-out by 2005, the fumigant was given "critical use exemptions"; and in California, the industry insisted that no feasible alternatives existed. As the twentieth century came to a close, over 35 million pounds of methyl bromide were being injected into US soils annually to protect berries and other crops including peppers, perennials, and tomatoes. About half of that was used in California. As the phase-out loomed, by 2004 growers had cut back to 7 million pounds.[14] By 2016, just half a million pounds of methyl bromide would be applied to strawberry fields, signaling the chemical's swan song.

Bereft of their go-to fumigant, berry growers and associated scientists like Lloyd have been scrambling for alternatives: a way to combat disease without resorting to the toxic chemical. The demise of this go-to fumigant may just turn out to be the necessity that will drive innovation. And the frantic search for replacements may have benefits beyond the strawberry field. One option is to seek out microbial allies.

Millions of Allies

Well before the twenty-first-century fascination with the microbiome, there was Sir Albert Howard, an English botanist and a pioneer of organic farming. Writing during the years when Dupont (and others) were filling our heads, homes, gardens, and agricultural fields with the promise of "Better Living through Chemistry," Howard advocated something different: compost, that magical black gold that we now know is thrumming with life and nutrients. If farmers cultivated healthy soils, he reasoned, they needn't be so reliant on chemicals. Essentially, Howard advocated for better living through nature. "In the nutrition of our population," he wrote, "at least two mistakes are made: the food is grown for the most part on poisoned soil; it is afterwards refined, processed, and preserved in various ways. Both these factors cause untold mischief."[15] He would likely be horrified by the foods lining our grocery shelves today.

Instead of chemical additives, Howard believed in the power of microbial life, writing: "This population of millions and millions of minute existences . . . come into being, grow, work, and die; they sometimes fight each other, win victories, or perish. . . . This lively and exciting life of the soil is the first thing that sets in motion the great Wheel of Life."[16] Compost—the agricultural equivalent of a fecal transplant—would leave soils teeming with microbes.

But soil that has been plowed, fumigated, and fertilized is a tough place for a robust microbial community to set up shop. In contrast to a newborn baby who carries her mother's immunity and a good bit of her microbiome, consider the tomato start, or the strawberry plant, or wheat seedling, which must begin anew each time it sends tender roots into the soil. These are plants grown to satisfy our palates and line growers' pockets, not to survive in nature. Many are stripped of their natural and often bitter-tasting defenses. They are plants in need of protection. When stuck in chemically treated soil, a plant recruiting a healthy rhizobiome is on par with a patient recovering from a bout with powerful antibiotics. Maybe they'll manage, or maybe an aggressive and opportunistic pathogen will take hold, or at least resist a chemical treatment.

In some fields, even as growers blasted soils year after year with fumigants, verticillium kept returning. So too did other microbes. Which begged a few questions: What goes on belowground? How did pathogens return? Where were these microbes hiding out? Were they hardier than their ancestors? Maybe sterilization wasn't the answer, but the problem. Maybe *Verticillium dahliae*, like *C. diff*, can take advantage of a microbial social order thrown out of whack.

Over half a century ago, Howard hoped to convert chemically dependent British growers into a "nation of compost gardeners," a practice that could help avoid this sort of invasion.[17] In their book *The Hidden Half of Nature*, Anne Bikle and David Montgomery write of Howard's proposed "salvage campaign to return vegetable, animal, and human wastes to the fields."[18] For Howard, the stakes couldn't be higher: the future of agriculture, and so, life as we know it. But compost was a tough sell when the government was subsidizing the fertilizer industry. Howard was a man before his time. Now though, with advances in technology, the timing is right. Genomics and metagenomics are providing scientists with a new vision for agriculture. A return to ecology that is better informed by technology. A return to traditional farming techniques. A return to the soil, where the emerging rhizobiome is understood as agriculture's "final frontier" and the key to the "second green revolution."[19]

Soil Is Not Just Dirt

Growing up in the suburbs of the sixties and seventies, I gave little thought to soil. Clothes got "soiled," but in the field next to my home, where emerald-green grasshoppers were easy targets for a girl with a net and a jar, the rye grass, Queen Anne's lace, and milkweed all set their roots in dirt. Dirt was what I dusted from the knees of my jeans and scrubbed from beneath my nails. It wasn't until I entered graduate school that I learned about soil. There, I discovered a world of soil microbes that jostled for space, gobbled up their smaller neighbors, and pushed out the more timid. These microscopic communities also produce a treasure trove of enzymes that can turn minerals into plant nutrients and chemicals that may act as specific, potent poisons. Aggressors and allies coexist in a constantly evolving ebb and flow of life and its associated chemistry. Dirt is inanimate, but soil is alive.

Growers know implicitly, if not explicitly, that soil is not just dirt. But how could they not be wooed by the power of pesticides and fertilizers to increase yields? In some cases, the same plot of land produces three or even four times what it did forty years ago.[20] Humanity has, no doubt, benefited from chemical inputs to farming, just as we have from antibiotics. Yet the chemically driven Green Revolution, combined with an ever-increasing number of hungry mouths, set off a "sow and sell" cycle that has left many agricultural soils in the sorry state lamented by Howard—devoid of nutrients and with microbial communities in disarray. Our willingness to throw "millions and millions of minute existences" under the chemical bus, along with other insults, has become worrisome enough that the United Nations declared 2015 as the International Year of Soils, as a reminder that soil is the "key to sustaining life on Earth."[21] Whether we call it dirt or soil, this mixture of minerals, carbon and other essential chemicals, and microbes is Earth's lifeblood. Lloyd and others know this, but convincing conventional growers to put their faith in soil rather than fumigants is a challenge.

One crop that gives scientists hope is wheat. Like most crops, the grain that becomes our daily bread (harvested by the hundreds of millions of tons globally[22]) is susceptible to plenty of pathogens: grass-green seed heads turn a dull brown with blight; smut turns them black; sooty molds leave seed heads looking spray-painted and spotty. Some pathogens drop in with the wind. Others lurk in the soils. Most insidious, perhaps, are the root rots: diseases that reach up from the soils, barely noticeable until patches of dead seed heads begin dotting the fields. Take-all disease, caused by a fungus (*Gaeumannomyces graminis*), blackens roots before making its way up to the crown of the wheat plant. Before long, ghostly white islands of diseased wheat emerge in fields of green. Worldwide, take-all is notorious. Sometimes the fungus takes half the crop, sometimes it takes it all. There are few cost-effective fungicides.[23] Rotating and fallowing (unseeded) fields have been among the few options available for wheat growers, and even then results are never guaranteed, not when a pathogen is lying in wait. But disease is not inevitable, and there is an ebb and flow to soilborne pathogens, just as there is with human epidemics. Sometimes a disease subsides because it kills off too many hosts, becoming a victim of its own success. Other times a host develops a stronger immune response and fends off the next wave. Or there is a change in the weather: a region becomes drier, or more humid. And sometimes, very small allies come to a host's rescue.

Early in the twentieth century, wheat farmers and scientists observed something curious. For years, they had watched crops grown on pathogen-infested land falter and fail under attack from take-all. Then one season, the crops didn't succumb. Take-all had declined, naturally. This unexpected waning of a notorious disease provided some of the first evidence that, even in a mono-cropped system, nature takes care of her own. (Note that unlike human populations that die off, leaving pathogens without a host, these fields were seeded again and again, providing the fungus with a steady supply of hosts ripe for infection.) The decline of disease went against dogma: how could crops planted in successive growing seasons in pathogen-infested soil begin to flourish, with no specific intervention?[24] The answer lies with soil microbes. The microbial community that once hosted take-all had changed; it was as if microbes had called in the peacekeepers to manage unruly citizens of the rhizosphere. Over time, the soil had begun to subdue the disease. These are disease-suppressive soils. And they are, says Lloyd, "absolutely the future." While take-all suppression can be cultivated over years of successive mono-crops, suppression of other diseases might be cultivated by rotating in crops that recruit helpful bacteria or discourage harmful bacteria. Disease-suppressive soils may even provide some relief for California's berry growers mourning the loss of methyl bromide.

Of Bacteria and Berries

You can hear the passion for berries, microbes, and growers in Margaret Lloyd's voice, even as it zings across 3,000 miles from her California office to mine in Massachusetts.

"Thinking about the complexities of commercial strawberry production," says Lloyd, "is one of my favorite things." Having spent her graduate years researching nonchemical replacements for methyl bromide, Lloyd (now a University of California small-farms advisor) is a fan of both berries and less toxic solutions. But, says Lloyd, "you have to have an alternative to methyl bromide that supports the future industry." Methyl bromide and chloropicrin not only blasted verticillium wilt away, but, for reasons no one has yet to adequately explain, the chemical pairing also boosted plant growth—which makes this killer combo a particularly tough act to follow. With an average California acre churning out forty to eighty *thousand* pounds of berries (three times more than Florida, and ten times more per acre than most other states), and the go-to solution fast becoming another industrial-age chemical relic, an alternative must both control pests and enhance growth.

There have been replacements, but they've had their own toxic baggage. Methyl iodide, a cousin of methyl bromide, enjoyed a few months of application before pesticide activists and growers, concerned about its toxicity, forced the supplier to retreat and pull it from the shelves. Another replacement turned out to be a potent greenhouse gas. Chloropicrin is one of the few remaining fumigant options, but it isn't as effective as methyl bromide when used alone. Simply ditching the chemicals and going all organic (even as the organic sector grows) isn't always a feasible option, especially for those who can't shell out seven or eight dollars per quart.

This is particularly true in California, where the cost of farming means growers are reluctant to switch. Land is pricy and prevention adds expense. In 2011, a conventional berry crop (managed by using Integrated Pest Management approaches), from roots to fruits, was estimated to cost a jaw-dropping $36,264 per acre.[25] For that price, berry plants had better make themselves useful. Lloyd, who would like to see growers move away from chemicals, says that raising organic strawberries in the Golden State can be done, but growing on a large scale is hard, and, on average, organic strawberries have a 30 percent lower yield. Organic growers, whether producing strawberries or other fruits and vegetables, have to grow smarter—integrating more-hygienic practices, farming more diligently, and monitoring for quick response. Like health care centers trying to stem the rise of hospital-acquired infections, growers need to farm preventatively. With a crop like strawberries, which must be tended to for over a year, pests, pathogens, and weeds are particularly difficult to control.

Not only do organic strawberry growers have to farm that much smarter, but they have to grow that much more. One standard practice for beating pest and pathogen is crop rotation, so strawberries are rotated to another plot for three or four years at a time—meaning more land devoted to the whole strawberry venture. As big-box retailers like Walmart fill their shelves with organically grown produce (for better or for worse, depending on your perspective), the acreage devoted to organics is growing—which means that more growers are becoming more rigorous in their practices. Those Red Fire Farm berries I picked at my local CSA? Pampered. At best, crops can stay in the same plot for two years, sometimes only one, says Red Fire owner, Ryan Voiland. "I generally try to plant strawberries into areas that had brassica crops the year before, and sometimes sorghum sudan."[26] There is evidence, says Voiland, that these kinds of crops can decrease levels of strawberry diseases and nematodes. And, as most any small-scale strawberry grower here in the eastern United States will attest, because berry plants are poor competitors, weeds are even more of a nuisance. Voiland's strawberry fields are cultivated almost weekly during the growing season—hoed and weeded by hand numerous times as plants get established. "They require more hours of weeding effort than almost any other crop we grow." All this adds to the cost of chemical-free production. For conventional growers in California planting acres of berries on expensive land, rotating with a crop like broccoli—not exactly America's favorite—is a tough sell. But this kind of rotation, especially with crops that can suppress verticillium, Lloyd says, is huge.

There are other emerging alternatives to chemical agriculture, says Lloyd, like pasteurizing soil with steam. Another method, called anaerobic soil disinfestation—suffocating life below ground (the oxygen-dependent species anyway)—is designed to slow disease, but these approaches dramatically alter the soil microbial community. Lloyd's heart and mind is with the microbes. "We are moving towards thinking about *managing* soil microbiology. With steam, for example, scientists are seeking to understand how long-lasting the disease suppression is," and, she says, how the treatment affects the soil's microbial community. Even with methyl bromide treatments, the microbes still remain somewhere in the soil. "The native microbial community," says Lloyd, "reemerges."

Managing soil microbes rather than blasting the hell out of them in order to kill a few is a radically different approach. This is where scientists like Dan Chellemi come in. His research with tomato crops in Florida and strawberry crops in Florida and California shows how rotational and cover crops can change the plant–host relationship: a shift in agricultural practices that tends to favor disease-suppressive microbes.[27] As a plant pathologist and former applied-research manager at Driscoll Strawberry Associates, Chellemi is a soil microbe evangelist. And he is on a mission to convert skeptics who might prefer to reach for a chemical cure while trying to keep their businesses afloat—to become both better businessmen and environmental stewards. Chellemi works with farmers, guiding them as they transition toward more-sustainable practices, including supporting the beneficial microbial communities in the soil. Natural disease-suppressive soils are "a reality," says Chellemi.[28]

As much as organic growers and soil-minded scientists like Sir Albert Howard appreciated the power of microbes in the fifties, sixties, and seventies, the tools they had available for studying life belowground were limiting. To farmers whose livelihoods depended on agricultural sciences, embracing the black box of soil microbes compared with the crystal-clear killing power of chemicals must have been difficult. "Now," says Chellemi, "there is a renewed interest in understanding soil microbiomes that is attributable to recent advances in metagenomics. It's very cool. . . . Before metagenomics, working with soil microbiomes was like watching a ballgame through a chain-link fence. You can kind of see what's going on, but you aren't on the field. Metagenomics puts us on the field. The black box is becoming less opaque, and this is a good thing."

If scientists like Chellemi and Lloyd are successful, soil microbes will once again be the key to sustainable agricultural systems. And growers who need to manage pathogens like *Verticillium dahliae* won't be seeking the next-best soil fumigant but asking how to encourage their soil's microbiome. It is a case of "back to the future," only now scientists are armed with molecular details unavailable to Howard. It's anyone's guess how quickly, if at all, conventional growers will begin cultivating their microbes. And, as with most complex problems, there won't be one solution that fits all farms. While Lloyd advocates for rotation with broccoli and other crops, scientists working with a fifteen-year strawberry monoculture in Korea recently reported disease-control success thanks to antimicrobials produced by disease-suppressive soils.[29] "Suppressive soils," write the authors, not only occur everywhere for just about everything, but they "provide a highly sustainable means to control pests using minimal off-farm inputs."[30] Like Chellemi says, naturally disease-suppressive soils work.

One Step Back and Two Steps Forward

Strawberries are but one crop among many. All told, about 150 different plant species feed the world, with a dozen major food crops made to bear the brunt of filling our bellies.[31] In 2007, US growers used over one *billion* pounds of pesticide; 5 billion pounds were used worldwide.[32] They not only linger in our favorite fruits and vegetables, but are a considerable expense to growers and to the environment. Many of these toxic chemicals are losing their edge as pests, pathogens, and weeds evolve under pressure to resist them, challenging growers to turn up the chemical volume or seek newer, more-powerful pesticides. The Americas combined use over 20 million tons of fertilizer. Globally, fertilizer use in 2018 is estimated to rise to over 220 million tons.[33] At the same time, the public increasingly demands foods that aren't all hopped up on synthetic fertilizers and are free from pesticide residues; in response, stores like Walmart and Target are increasingly stocking organics along with their so-called conventionals.

Twentieth-century chemistry turned soil into dirt, but twenty-first-century technologies like genetic sequencing and metagenomics, combined with good old-fashioned farming practices, can help us turn it back into soil. Even as knowledge of the soil microbiome and disease-suppressive soils grows, figuring out how to cultivate soils to best serve crops will require time, money, and devoted scientists. Even then, suppressive soils won't be the answer for every pest and pathogen in every crop. Likewise, fecal transplants work for *C. difficile* in some cases, but transplants won't cure all patients or all intestinal ailments.

Growing food is a complex and often risky venture made even more difficult by the constant onslaught of pest and pathogen. The extent to which industrial growers will embrace life belowground as they cultivate life above isn't yet clear. Nor are beneficial microbes a panacea. We might encourage hearty microbiomes in our own bodies, but there will be times when we will still have to turn to antimicrobials to fight infection. And on the farm, we will still need pesticides, whether chemical or biological. But what if those cures and chemicals were "smarter?" What if a pesticide *targeted* wilt, or black rot, or aphids? Or recruited helpful allies? What if cures for staph and strep acted like predators zeroing in on their prey? These solutions exist. Some seem like cures from a science-fiction future, others will leave us scratching our heads as plants push pests away with chemical trickery. Either way, culturing microbiomes is a fundamental step in the right direction. Next, we turn to cultivating the enemy of our enemies.

CHAPTER 3

The Enemy of Our Enemy Is Our Friend: Infecting the Infection

A TODDLER SUDDENLY BECOMES DEATHLY ILL. In the ER she is diagnosed with dysentery, caused by a rare but particularly aggressive form of *Salmonella.* One antibiotic after another fails because the strain, picked up when her family was traveling across Asia, resists multiple antibiotics; but there is an alternative new drug. Like a guided missile, the drug targets this particular strain of *Salmonella.* Not only that, but as long as *Salmonella* remains, the drug particles replicate, increasing in number until the infection subsides. Despite the carnage, the toddler's gut microbiome remains unharmed—no need for probiotics or fear of complications like *C. diff.* If the *Salmonella* responds by evolving resistance, the drug may respond in turn, engaging in an ages-old evolutionary dance. By the next morning, the color returns to her cheeks. By evening, she is cured.

While still a fantasy here in the United States, scenarios like this have been playing out in Eastern European hospitals and clinics for nearly a century. The "new" drug is a cocktail of viruses called bacteriophages (known simply "phage therapy") that attack bacteria. It is a cure nearly as old as life—at least as old as bacteria. Microbiologists have suggested that for every strain of bacteria on Earth, from the oceans to those populating our own microbiomes, there is at least one, if not multiple bacteriophages. We are fast becoming desperate for new antimicrobials that are both effective and that cause minimal damage to our own microbiomes. Bacteriophages are potent antimicrobials. Once disparaged here in the United States and in Western medicine in general, these bacteria-infecting viruses are making their way back into academic and biotech laboratories. If all goes well, they may be coming to a pharmacy near you.[1]

It's Only a Phage

While we have been slow to recognize the benefits of bacteria, we are even farther behind when it comes to viruses. Essentially tiny bits of genetic material wrapped in protein, viruses are so minuscule that they slipped through the early ceramic filters designed to isolate bacteria. And so as nineteenth-century medical scientists began identifying one pathogenic bacterium after another, viruses remained undiscovered, indistinguishable from any of the common molecules or particles in lab solutions—except that these "particles" could replicate and seemingly thrived on living tissues. Like spirits, they were invisible, enigmatic, and potentially deadly. Despite the medical and technological advances that were made throughout the golden age of microbiology, viruses continued to cause havoc. One of the most notorious outbreaks was the great influenza pandemic of 1918, which killed tens of millions of men, women, and children—possibly over a hundred million.

What exactly these infectious particles were—animal, vegetable, mineral, or none of the above—would remain a mystery for decades. "When viruses were found to possess the ability to reproduce and to mutate, there was a definite tendency to regard them as very small living organisms," wrote Wendell Stanley upon receiving the 1935 Nobel

Prize for his role in revealing their nature.[2] Yet even Stanley was initially reluctant to acknowledge that viruses were not just bits of protein, but also, at their core, contained some genetic material.[3]

It is that bit of genetic material that makes them so relevant to our lives. Eighty years later, we know that Earth is teeming with viruses, some causing us to sniffle and sneeze, others killing within hours. We know that they are parasites: they not only invade a cell but use the cell's machinery to essentially clone themselves, sometimes leaving the infected cell in tatters as hundreds of new viruses emerge, sometimes leaving bits of their own genetic material behind, embedded within the host's own genome. Most of these bits are harmless, some are helpful, and a relative few are harmful. We now know there are a billion viruses in a teaspoon of ocean water and trillions living within us.[4] And we know that, throughout our existence, viruses have woven in and out of life—leaving their stamp on most if not all living things. By some accounts, up to 8 percent of our genetic material came to us by way of viruses.[5]

Yet, for all the fear and harm we associate with viruses, many (if not most) are phages, infecting bacteria, like those in our microbiome. Genomics is just beginning to reveal the diversity and representations of these entities in nature and within our bodies.[6] But the role that phages can serve as potent antimicrobials is no mystery. As infectious agents of bacteria, they are a normal and pervasive component of Earth's flora. When directed toward human pathogens they can save lives. One day they just might save us or our loved ones—as they did more than half a century ago when phage treatment saved Gary Schoolnik's mother. Schoolnik, an emeritus professor of medicine at Stanford University with a specialty in infectious disease, tells the story of his mother as she was dying from typhoid fever in 1948. At the time there were no antibiotics effective against the particular *Salmonella* bacterium that caused the disease. His father, a Seattle surgeon, had read about some doctors in

Los Angeles using bacteriophage to cure typhoid. Desperate for a cure, Schoolnik's father contacted the group, acquired the phage, and injected his wife. Within two days, her fever disappeared.[7]

This was a time when antibiotics were becoming the symbol of modern medical success, but typhoid was still one of the last holdouts. When an effective antibiotic was finally discovered, any enthusiasm for phage quickly waned. The therapy also suffered from conflicting reports about efficacy, poorly controlled clinical studies, and a lack of quality control by companies producing phage. As a result, the science of healing with viruses, writes Anna Kuchment, in her aptly titled book *The Forgotten Cure*, "had been all but forgotten in the West." Now, though, Western scientists and physicians are trying to bring the therapy back to the American pharmacopeia, but doing so means navigating what has—for good reason—become a rigorous and expensive drug approval process under the US Food and Drug Administration (FDA). This process, though, is also relatively rigid, so that even *if* phages were to pass efficacy and safety trials with flying colors, roadblocks to approval would remain.

The drug-testing gauntlet set up by our FDA was designed to test chemicals, with dosages based on chemical concentrations, taking into consideration absorption, distribution, excretion, and the safety of specific molecular structures. FDA approval is thus also based on well-defined dosages and known chemical makeup. But phages don't work that way. They may be administered in huge numbers and in combination cocktails, depending on the infection—*and* they may replicate within their host. This explains, in part, why phages are not yet an option for treatment of human disease here in the United States. And why, rather than more stories like Mrs. Schoolnik's, instead we have more recent tales of desperate patients like Laura Roberts, a Texas mother who suffered from a chronic and nearly lethal infection caused by methicillin-resistant *Staphylococcus aureus* (MRSA). In 2006, after being told that it was time to get her affairs in order, Roberts, who has shared her story in books and magazines, as well as on CNN, traveled to the Phage Therapy Center in Tbilisi, Georgia, a world-renowned center for developing the therapy.[8] As she tells it, she made the trip using a walker and with her brother for support, her body frail and under attack. After three weeks of treatment, including phages specific for her MRSA strains, Roberts left for home—sans walker.[9] While the combination of treatments Roberts received makes it difficult to give the phage mixture all the credit, it likely played a significant part in curing her.[10]

Shigellosis—a severe form of bacterial dysentery—was the first human disease targeted with phage therapy. Working around the turn of the twentieth century, microbiologists Félix d'Herelle, a French Canadian, and Frederick Twort, an Englishman, discovered (independently of each other) an invisible agent that made bacteria magically disappear. D'Herelle continued to study the oddity, eventually pitting the mysterious cure against dysentery. After ensuring that the therapy was safe by ingesting phage (he did so along with other willing participants), d'Herelle cured his first young patient, a twelve-year-old suffering from severe dysentery, with phage collected from the feces of infected soldiers. The results were miraculous. The fever and characteristic bloody stools subsided within twenty-four hours.[11] He repeated the treatment on three other desperate young patients, with similar success. Within a decade, phage therapies, packaged into oral, injectable, and topical medications, were treating infections due to *Staphylococcus*, *Streptococcus*, and cholera bacteria, with varying degrees of success. Phage cocktails, or mixtures, ensured that despite (or, in some cases because of) the phage's specificity and the lack of rapid accurate diagnosis, treatments would be effective. Pharmaceutical companies on both sides of the Atlantic—including Abbott Labs, Squibb, and Eli Lilly here in the United States—joined in, developing various phage therapies of their own. D'Herelle's work was recognized with multiple nominations for the Nobel Prize in Medicine.[12] But then came antibiotics. And the Cold War. And, it should be noted, poor practices by some pharmaceutical companies: commercial products in the United States were often found to be lacking in potency. Finally, a couple of damning reviews in the esteemed *Journal of the American Medical Association* helped to close the door on phage therapy in Western medicine.[13]

Meanwhile, Eastern European countries continued to refine the therapy. Researchers understood that bacteria evolve resistance under pressure from killer viruses, but viruses evolve even faster. Phage cocktails could be updated as newly resistant bacterial strains emerged. Today, in at least some parts of the world, such treatments continue to heal. Elizabeth Kutter, a professor at Evergreen State College in Olympia, Washington, has devoted her career to the study of bacteriophages. Speaking about phage therapy at the International Conference on Biotherapy in 2010, Kutter assured the audience that, yes, "it's a real thing." The therapy, says Kutter, is targeted, self-replicating, can penetrate more deeply into infections than antibiotics, can evolve, and can even behave as an "infectious cure," spreading from one infected animal to another.[14] After working on basic bacteriophage research throughout the seventies and eighties, Kutter was first introduced to phage treatment in what was still, in 1990, the Soviet Union. At first, recalls Kutter, she was suspicious. "I thought, you know, if nobody is using it anywhere and if I've never heard of it, how could it be true?" Kutter has since become a proponent, devoting her laboratory to phage therapy research, hosting conferences on the topic, and collaborating with research organizations. Today, more-powerful and less-expensive DNA sequencing means that infections can be more accurately diagnosed, and their associated phages more readily identified. Phage "banks" provide an opportunity for responding to infections with individualized cocktails.[15]

Yet, without an approved therapy here in the United States, one version or another of Laura Roberts's story continues to play out, with patients traveling to places like the Phage Therapy Center in Tbilisi for a cure. In 2015, out of options and desperate, Suzanna, a forty-three-year-old Chicago mother, took the plunge and, like Roberts, traveled across the Atlantic. For five years she had been suffering from incessant sinus and bronchial infections; a slew of different antibiotics provided no relief. Surgery didn't help either. "I was given hope that all would heal," she says of the surgical option, but after surgery the symptoms not only persisted but grew worse over time.[16] Two years later she was diagnosed with MRSA, but doctors had no good suggestions about how to overcome it. Miserable from her symptoms in addition to having developed allergies to many medications and foods, Suzanna says she had nearly given up when her husband came to the rescue: "My husband found Chris Smith [founder of Phage Therapy International] and the Phage Therapy Center in Tbilisi. I was skeptical but willing to try anything to feel better and overcome MRSA."

After doing some due diligence, Suzanna decided she had nothing to lose by giving the therapy a try. For ten days, she shuttled between her hotel room and, across the street, the Phage Therapy Center, where she was subjected to a combination of therapies, including phage therapy and also immune therapy to boost her immune response, that were designed to treat MRSA and root out the rampant candidiasis she had developed after years on antibiotics. (*Candida* yeasts, like *C. diff*, are a notoriously opportunistic group of microbes that can grow out of control in response to antibacterial treatment.) She was also treated for *Klebsiella*, another microbe that had apparently slipped past her own doctors in the United States. Each day, she spent two hours at the clinic and then rested at the hotel, where Suzanna says she "drank fluids and inhaled or nebulized with phages." Within days, she was feeling better. By day nine, both her sinus and any intestinal problems she'd been having cleared up. "I came back to the US and surprised my doctor with my general appearance," she says. "We then tested again three months later to find that the MRSA and *Klebsiella* were nonexistent. . . . It is a story of a second chance. My entire life turned around after being made well."

Stories like Laura's and Suzanna's are hard to ignore, even for practitioners of traditional Western medicine, or drug developers, or regulators. In 2013, the European Commission funded a large study testing phage against skin infections that commonly plague burn patients.[17] Results are expected in 2017.[18] In 2016, AmpliPhi Biosciences, a biomedical firm headquartered in San Diego, California, that is just one of many companies hoping to bring phage therapies to market, reported that they had a couple of phage products in the early phases of clinical trials. And in June of 2015, Randall Kincaid, senior scientific officer at the National Institutes of Health, organized a meeting of international bacteriophage scientists, entrepreneurs, and regulators hailing from the United States, France, Georgia, China, and elsewhere.[19] Ever since heading up a US Department of Defense program in 2009 to counteract possible attacks with engineered bacteria, Kincaid has been thinking about phages. "There are ten- to fifty-fold more phages than bacteria in nature. They are very successful. If there are concerns about (engineered) anthrax, then there must be a collection of phages that can attack it." So what's the holdup here in the United States and other Western countries? In a word, research.

"The conclusion from the conference," says Kincaid, "is largely that many treatment effects are poorly documented"—at least in Western terms. Here in the United States, drug testing, particularly once it moves into human trials, requires randomized, blind studies where physicians don't know whether they are providing the test drug or a drug that is already approved. Kincaid explains that in the former Soviet Republic of Georgia, and in Poland, the studies are not done this way. "Not that they are not credible," he says, "but they didn't apply the same standards to evaluating the treatment . . . so there is skepticism that they would be broadly useful." That is one key to turning phage into drugs: demonstrating that they are effective in treating an infection again and again and again. But such skepticism may be quelled with sufficient attention: more funding, testing, and interest in developing phage therapy. And the growing problem of antibiotic-resistant bacteria may well be the factor that will move phage from Eastern Europe to a hospital near you.

Doctors at Tbilisi's Phage Therapy Center were able to treat Laura Roberts's and Suzanna's infections with a customized phage mixture, but this kind of customization doesn't yet fit into the FDA framework. "This is an ultimate view of precision medicine," says Kincaid, "but from a product point of view and a regulatory point of view, that is a much different way of looking at the world." This wouldn't be the first odd duck that the FDA must regulate. Because influenza viruses are notorious for evolving, each year we are offered a new flu vaccine. The active ingredients are viruses (either inactivated, live, or fragmented). New combinations are prepared, depending on the prevailing and predicted flu strains. For decades, the FDA has worked with vaccine producers to ensure safe, effective vaccines. How this model might apply to a product like phage is anyone's guess. But like flu vaccines, phage therapy may require some modification of the cocktail over time in order to "adapt" to bacterial resistance. As phage-therapy products make their way through the testing and clinical trial phase here in the United States, developers will soon be knocking on FDA's door. (Testing aside, there are other issues, too, like intellectual property rights. Nature cannot be patented, and while patent applications for phages are piling up here in the United States, it is unclear how these therapies will be handled by American patent offices. If pharmaceutical companies don't own the exclusive rights to manufacture brand-name phages, it could deter investment in research. Perhaps in this case, need will win out over profits.)

While a drug must work it must also be relatively safe. No one wants another thalidomide—the sleep aid that caused an epidemic of deformed limbs in newborns in the 1960s; it was approved elsewhere in the world, but not the United States. Phages, though, are not synthetic drugs, nor are they human pathogens. Our bodies are already teeming with phages. Many are engaged in an age-old battle with bacteria in our guts, our mouths, or on our skin, helping to keep dominant and beneficial species in check, and ensuring microbial diversity.[20] If we are squeamish about being exposed to a live virus *intentionally*, we would do well to recall that we've been ingesting them for decades in processed meats and cheeses, in addition to whatever viruses we encounter naturally. Phages used for infecting food-borne pathogens like *Listeria* and *Salmonella* have been "generally recognized as safe" (that is, "GRAS") by the US Food and Drug Administration and other regulatory bodies around the world for years. And, of course, they are 100 percent organic.[21] In fact, says Kincaid, "what it says to me is that in other parts of world, phages have the ability to solve other types of problems, perhaps in some ways more important to human health initially than the actual treatment of disease."

Still, until phage therapy passes regulatory muster, Western physicians will continue to yearn for the treatment, perhaps ordering phages from afar and using them at their own risk. And patients in need will suffer the consequences, whether from a treatment in need of some regulatory oversight or for *lack* of a treatment that could save their life.[22] As we embrace our microbiomes, we are increasingly trying to understand how complex societies of bacteria, fungi, viruses, and other microbes interact. How can one strain or species be controlled, without indiscriminate destruction? Bacteriophages are one way. But there are others. "Nature," says Kincaid, "has provided us with a great deal. We could try and leverage that."

Chemical Warfare, Naturally

Another treatment to emerge from nature's bounty is a group of highly specific killer proteins called bacteriocins. University of Massachusetts microbial ecologist Margaret (Peg) Riley has studied bacteriocins for most of her thirty years in research. Much of her work has focused on ecology and evolution. But now, like Elizabeth Kutter, Riley is on a medical mission. Her goal is to infuse some ecology into disease treatment, to turn bacteriocins into medicine. "We are very ignorant of the biology of the microbial world," she says, but with new approaches such as metagenomics and microbiome studies, "we are getting glimmers of a whole world full of wonder" and a world potentially full of cures.[23] These small proteins, produced by the vast majority of microbes and active against closely related microbes, promise targeted destruction of infectious bacteria. Bacteriocins are like weapons in a fratricidal war. One strain of *E. coli* against another strain of *E. coli*. *Salmonella* against *Salmonella*. Like phages, their power lies in their specificity: they are good for the microbiome and good for slowing the evolution of resistant bacteria.

"All of my early work had nothing to do with therapeutics," explains Riley who, like a kid bursting with an idea, exuberantly turns to the whiteboard in her university office and begins sketching a diagram of how bacteriocins kill. What makes bacteriocins so specific and potent, she says, is not *just* the protein, but the ability of the cell on the receiving end to recognize the protein. It's like a futuristic lock that attracts a key magnetically, and once the door is unlocked—that is, once the bacteriocin begins interacting with receptor proteins on the cell surface, it begins to cause structural changes. Pointing to the diagram, Riley says, "this killing bit right here can enter and form a pore and all the guts pour out." Game over. "These are *cool* proteins," she enthuses, and when Riley recalls her earlier encounters with bacteriocins, her eyes light up: "Wow, it looks like microbes are in constant warfare. I just thought that was amazing." Eventually, a colleague suggested they might try to turn these bacterial weapons into therapeutics: an antidote to the menace of antibiotic resistance. But the work was slow going. The benefits of specificity notwithstanding, the market for a drug that treats only MRSA or only tuberculosis would be very narrow. There were other issues as well. Proteins can be metabolized in the body and so put out of commission. They may have a relatively short shelf life and, in many cases, would have to be injected. But now, with the very real threat of resistance, the new focus on maintaining the integrity of our microbiome, and a host of technological advances, the playing field has changed.

Earlier in her career, Riley and coworkers demonstrated how bacteriocins interact with their bacterial targets and how they promote intraspecies ("within-species") diversity. Unlike the phage, which essentially clones itself once it attacks, microbes that produce bacteriocins make the ultimate sacrifice. "If you are carrying a bacteriocin, you die when you produce it. It is a very lethal weapon just to carry on your body. You lyse open [break apart] and release the bacteriocins. I think of it as form of bacterial altruism." Cells that are *closely* related (bacteria reproduce mainly by cloning) will survive because they have genes for immunity. Any other strain will die, save for those very few that evolve resistance. But there is a cost to producing the deadly proteins and for becoming resistant; the result is a rock-paper-scissors kind of dynamic. A producer strain kills the sensitive strain; the sensitive strain outcompetes the resistant strain; and the resistant strain beats the bacteriocin producer.[24] Eventually the population reverts back to a sensitive strain. It's an uneasy balance that ensures diversity rather than the emergence of a single victorious strain. This complex dynamic plays out wherever bacterial communities exist, from the rhizobiome of plants to our guts, where coliforms, *Salmonella*, and others are probably lobbing bacteriocins at one another as you read. As with phages, we have a very long history of bacteriocin exposure.

Riley isn't the only one singing the praises of bacteriocins. One bacteriocin in particular, nisin, has been added to our foods for decades. Produced by strains of *Lactococcus lactis* bacteria, nisin is a common inhabitant of naturally fermented foods like pickles and yogurts. Unlike many other bacteriocins, it is broadly active. This property, in combination with its "GRAS" designation by the US FDA, has turned the little protein into a modern-day food-preservation rock star.[25] And well before that, for thousands of years our ancestors have unwittingly been in cahoots with *Lactococcus* and other fermentation bacteria. Lactobacilli on cabbages and cucumbers and in milk products churned out nisin that killed harmful bacteria like *Clostridium botulinum* and *Listeria monocytogenes* (both of which can cause fatal infections, and in more modern times have triggered food recalls).[26] It is an effective and relatively harmless little protein, which makes it attractive to companies like Immucell in Portland, Maine. Immucell, with whom Riley now collaborates on some bacteriocin-related products, is hoping that theirs will be the first bacteriocin-based medicine to be approved by the FDA, albeit as a drug for veterinary use. Their goal is to provide a cure for mastitis, a common infection in dairy cows.[27] Riley hopes that approval of a bacteriocin for veterinary use will open doors for similar drugs in human medicine.

Each year, more than 20 billion gallons of milk are produced here in the United States. But our milk, cheese, and ice cream rely upon an antibiotic-heavy industry, in part because of mastitis—a painful udder infection that can be caused by a constellation of bacteria, including species of strep, staph and coliform, and that costs the industry billions of dollars. At worst, farmers cull chronically infected cows. At best, cows suffer from low-level infections that reduce milk production and quality. Somewhere in the middle, milk is discarded by the tankful. Infected cows (including those suspected of infection but not yet diagnosed) are treated with a handful of FDA-approved antibiotics, most of them belonging to a group of antibiotics called beta-lactams. If that term sounds familiar, that's because drugs like penicillin, amoxicillin (which is approved for use against mastitis in cows), and Keflex, all part of our human pharmacopeia, belong to the same class of drugs. Which means that it's even more important to reduce their use on the farm.[28] To protect consumers from direct exposure (treatments are often administered intramammary—right into the mammary gland), milk from treated cows is discarded until the presence of antibiotics drops to acceptable levels, adding to the cost of the disease. Improved hygiene and vaccination can help. But when prevention isn't enough, Immucell is betting on nisin as an alternative. Aside from taking antibiotics like beta-lactams out of the equation, nisin's GRAS designation, along with its effectiveness at very low concentrations, means farmers would no longer have to toss out all that milk.

Meanwhile Riley, along with recent graduate student Sandra Roy, is targeting human urinary tract infections, or UTIs, with a bacteriocin other than nisin. Hundreds of millions of new UTI infections occur each year, mostly in women, and they are notoriously antibiotic-resistant.[29] UTIs associated with catheterization are the most commonly reported health-care-associated infections. According to the Centers for Disease Control and Prevention (CDC), 15–25 percent of hospital patients will be catheterized during their stay.[30] As a catheter wends its way into the urinary tract, it may pick up microbes normally associated with a patient's bowels—an unfortunate consequence of our anatomy. Most healthy people won't be bothered, but others, depending on their condition and how long they have to use a catheter, may become infected. If the invading bacteria form a biofilm (a dense, nearly impenetrable layer that can build up on devices like catheters), they can be particularly difficult to reach with antimicrobials. Riley and Roy hope to reduce the chances of such infections occurring. One of the most common causes of UTI is *E. coli*, which also happens to produce one of the best-studied and the oldest-known class of bacteriocins: colicins.

Unlike nisin, colicins are highly specific, active against certain strains of coliform bacteria. Both can penetrate biofilms. But as with phage therapy, large-scale testing and eventual approval of bacteriocin-based drugs is expensive and daunting. So Riley and Roy are taking a different tack. "We can create a *lubricant* that can eliminate catheter-acquired UTI. So that's what we are going to go after first." Getting approval for new medical "devices," like a lubricant for use with catheters, is far easier. UTIs may not be a sexy first target for a potentially groundbreaking advance, but Riley and colleagues have even bigger plans. What if bacteriocins—even one specific for *E. coli*, like colicin—could be directed toward more-intractable infections like MRSA or tuberculosis?

About a year or so ago, says Riley, a Chinese researcher named Xiao-Qing Qiu contacted her about reviewing a paper he was trying to get published. For the previous fifteen years, Qiu had been working with colicins, trying to make them into medically useful antimicrobials. "He was interested in MRSA and noticed some work on the discovery of pheromones in bacteria. Staph was one genus for which a lot of work was done on pheromones." Bacterial pheromones are proteins that cells produce when populations get dense. As a form of intercellular communication, pheromones can induce populations to react. They may cause cells to form a biofilm, or to move away, or to produce other chemicals. Like bacteriocins, they are species-specific. Qiu's idea was to fuse together a well-known but very specific killer, colicin, with pheromones for disease-causing bacteria like MRSA. "Lo and behold," exclaims Riley, as if she were just now witnessing it anew, "it worked!" Not only did the new *pheromonicin* kill MRSA in animal tests, but was harmless when mixed up in a test tube with other kinds of bacteria.[31] Unlike the nisin that is directed against mastitis or used elsewhere, however, pheromonicins aren't all-natural, requiring a bit of genetic engineering to link bacteriocin to pheromone. Despite the implications of a potentially powerful and directed drug, Qiu's publications barely made a ripple here in the United States. "Probably because he is a Chinese scientist," says Riley. "Unfortunately, there are huge problems with being a Chinese scientist." It probably didn't help that Qiu was accused of fraud by other Chinese scientists.[32] In the end, though, as reported in the journal *Science*, Qiu was cleared of any wrongdoing.[33] A decade later, after continuing with his studies, he reached out to Riley, who not only reviewed his papers but has become a collaborator, testing his drugs in her laboratory. "I thought it [the work] was extraordinary. I have confidence that what he is publishing is real data." Riley describes the whole thing as *almost* too good to be true: "Until we actually have a drug out to market, I'm still going to think it's still a fairy tale." Working with Qiu has brought Riley closer to her goal of developing "designer drugs": highly specific, safe, and administered with lessons learned from our antibiotic-resistance debacle.

For nearly a century, we have managed bacterial infections with traditional antibiotics. Penicillin, streptomycin, and vancomycin are pitted against all-too-common infections, from pneumonia to staph to tuberculosis. Strains of these pathogens now resist multiple antibiotics. Over the course of a year, 2 million Americans become infected with resistant bacteria. Tens of thousands die as a result, according to the Center for Disease Control's annual statistics. Yet discoveries of new antibiotics have been rare enough that when molecular microbiologist Kim Lewis reported a new method of seeking out antibiotic-producing bacteria, news of his discovery went viral. But even if we do manage to enter into a new age of antibiotic discovery, scientists and doctors agree that we need not only effective antimicrobials, but also antimicrobials that act more like targeted drones rather than cluster bombs. A more potent and specific killer is a more difficult weapon to resist.

Resistance aside, there are other reasons to become more respectful of our antibiotics; we are so accustomed to the drugs that we forget how powerful they are. According to the CDC, antibiotics are the most common cause of medication-related pediatric emergency visits.[34] In addition, antibiotics may also contribute to serious chronic side effects such as diabetes, asthma, and some inflammatory bowel diseases, though a causative link has not yet been proven.[35] And then there are the opportunists like *C. diff.*

Both phage and bacteriocin could open up a whole new cornucopia of drugs. But they must first make their way through the long road to FDA approval before entering the doctor's office or our medicine cabinets. Some new drugs—particularly those that exist in nature—may make it more rapidly than others. Others may require some tinkering: removing or adding bits of protein to make them more effective, longer lasting, or better able to reach their targets. Phages and bacteriocins probably will not replace antibiotics even when they are approved, but they will certainly bolster our arsenal against infections.

And once they do enter the drug pipeline, we'll do well to learn from the past. For eons, nature has kept life in check, from the largest predator to the smallest. As we learn better how to live with microbes, we will increasingly be looking to nature as a collaborator. Phage and bacteriocins are powerful examples, in medicine and increasingly in agriculture. These are biological solutions that may target a particular problem without causing another. Our lives may one day depend upon them.

As difficult as human infectious diseases are, they are *relatively* simple compared with the vast array of pests and pathogens on the farm. Consider this: we are a single species infected by microbes that have adapted to thrive in and upon our body. But those who grow the wheat for our daily bread and the fruit and vegetables for our table must contend not only with pathogens adapted to infect a diversity of species, but also a mind-boggling number of insects and weedy pests. Some favor apples, others corn or cassava or potatoes. In recent decades, broadly acting pesticides have become the obvious answer, but they carry enormous drawbacks. And so, like medical researchers, innovative farmers are now seeking natural alternatives to synthetic chemicals.

CHAPTER 4

The Enemy of Our Enemy Is Our Friend: Replacing Pesticides with Nature's Chemistry

THE BACKYARD VEGETABLE GARDEN was a disaster. Deer grazed the peas, squirrels nipped the buds off the squash plants, and blight took the tomatoes. All that remained by August were the grapes, small green bunches of promise, until those too came under threat. Seemingly overnight, caramel-colored spots appeared on the ripening fruits. Within days whole bunches were infected. If they didn't drop from their stems, they shriveled into bitter, blackened raisins. This was black rot: a fungus that could take down a healthy-looking crop within a few days.

Determined to salvage *something* from the garden, I purchased a bottle of Serenade. This is bacteria in a bottle, a biological pesticide comprised of *Bacillus subtilis*, a common soil microbe. The bacterium produces powerful antimicrobial chemicals, useful in the wild for keeping neighboring bacteria in check. I sprayed with a vengeance, maybe even a bit more than recommended, but this was war. I'll never know if it was the Serenade, a change in the weather, drastic pruning, or, more likely, some combination, but by late August, red, blush, and pale green grapes dangled from the trellis, translucent in the morning light. Serenade and other biological treatments, from pheromones to predatory insects, are becoming a hot commodity, not just for the backyard gardener but even for large-scale growers. All-natural compost, crop rotations, and healthy rhizobiomes offer some respite and protection, but they can only take a farmer so far. When a fungus settles on ripening apples, or globetrotting weeds steal nutrients and real estate from hardy wheat or corn, even a robust soil microbiome can use some help. Biologicals are another tool in nature's arsenal.

Bacteria in a Bottle

The bacterium *Bacillus thuringiensis* (Bt) was first discovered in 1901 when it started attacking Japanese silkworms. Two decades later, rediscovered in the gut of the flour moth, Bt became one of the first commercial biological pesticides. Packaged in France as Sporine in the 1930s, Bt bacteria and the crystalline Bt toxin they produce while forming spores have been deployed in one formulation or another ever since. The dormant spore stage enables organisms like Bt to weather unfavorable conditions. Some spore forms can last for centuries. Commercial spores only need to remain viable for as long as it takes to travel from manufacturer to shelf to end use. Sporine targeted flour moths. But other toxins, produced by different *Bacillus* strains, are active against specific insect larvae, from moths to beetles and even mosquitoes. For over half a century, spore-forming bacteria, dusted, sprayed, and delivered in flake form, have provided some respite from pests on crops and in homes and gardens. Of the nearly 200 Bt products now registered as pesticides, several are approved for use by organic growers.[1] It is a good bet you can find plenty of Bt at your local Agway or Home Depot. (Though now a hot-button subject in the GMO debate, the gene for Bt toxin has been engineered into crops like corn and soy to protect against pests such as the corn root worm. More on this later.) The fact that this one species has been so beneficial to growers suggests that there must surely be others available for natural pest control. And there are.

Grown in large vats, sprayed on a field, or delivered as a seed coating, bacteria are living chemical factories. Advances in genomics and analytical chemistry are promising a whole new trove of microbe-based pesticides. The problem is in the packaging. Spore formers, like military MREs (meals, ready-to-eat), come in their own resilient packet that can last for years or more, ready to go when you are. The Serenade that I used contained a spore forming *Bacillus*, happy to perform wherever and whenever. Most bacteria, however, simply can't be bottled and released and expected to be full of vigor. They need to be fed and watered, they require just the right temperature, and like most living things, they can be finicky. Large-scale use of active bacteria is less attractive than spore formers on the farm . . . unless there is a way to kick-start growth, to whip a bacterial culture into a reproductive frenzy. One company, 3Bar Biologics, says they can do this: bottle up and deliver hordes of non-spore-forming bacteria (specifically, *Pseudomonas fluorescens*) that are raring to go.

Pseudomonas fluorescens (or *P. fluorescens*) is a bioprospector's dream: a battle-ready microbe that is an adept colonizer, gaining and occupying territory near plant roots, squeezing out less aggressive competitors, and sucking up nutrients. Once colonized, a plant can ramp up its own systemic resistance—much like our own immune system—to ward off infection by other microbes. Like many of its cousins, *P. fluorescens* is also a microscopic weapons factory, producing chemical compounds and proteins capable of busting apart or poisoning competitors while encouraging plant growth. One of those chemicals, 2,4-DAPG, is effective against a diversity of plant-disease-causing fungi, bacteria, and worms.

"Back in the 1990s," says microbial ecologist Brian McSpadden

Gardener, "Monsanto looked at DAPG, but abandoned it for lack of chemical stability . . . and most companies stayed away from *P. fluorescens* because it didn't have a long shelf-life."[2] McSpadden Gardener, a Midwesterner and self-described "agronomist at heart" believes that feeding the world organically isn't an impossible dream, and he's spent his career helping growers realize this. "If we can see the microbiome on the plant," says McSpadden Gardener, "then we can find which things are causing problems and which are helping." Identifying the beneficial microbial communities can lead to strategies on the farm that nurture soil health so that plants grow better, remain healthier, and are less dependent on a diet of traditional farm chemicals. While working at Ohio State University, McSpadden Gardener discovered a set of particularly useful strains of *P. fluorescens*. But there was that shelf-life issue. And so, in 2013, he cofounded 3Bar Biologics along with Bruce Caldwell to solve the problem. They say they've done it.

3Bar packages living *P. fluorescens* in bottles that can be turned into mini-bioreactors with the push of a button. 3Bar's bacteria are maintained in optimal conditions in 3Bar's own space, alive and ready to grow. Just before applying the product to their crops, growers release bacteria into a liquid growth medication. Like baking yeast when treated to a bit of warm water and sugar, the population explodes. Growers can inoculate corn and soy fields with hordes of active bacteria. So far, corn and soy growers are reporting increased yield: on average 4–5 percent—enough to keep growers wanting more. While spore-forming microbes like *Bacillus* can survive in a warehouse, says Caldwell, "if you limited yourself to spore-formers you limit capability. There are thousands of research articles about beneficial bacterial species . . . but there really aren't many effective commercial products out there."[3] The technology may someday help bring other beneficial microbes to the field. Bolstered by demands for sustainably grown produce, microbial bioproducts like those offered by 3Bar are becoming more widely used. Says Caldwell,

"I thought organic growers would be the heart of our user base; that turned out not to be the case. The vast majority that have tried the product are conventional growers."

A growing number of businesses are looking to profit by packing and selling all-natural defenses for pest and pathogen. Monsanto, Bayer Crop Science, BASF, and just about every other major agrochemical corporation is now dipping into the biological trough. In 2014, Monsanto claimed to have tested hundreds of microbial strains on over 100,000 farm-field plots, with a goal of doubling the number in 2015.[4] At a recent meeting about the problems of pesticide resistance, with an audience whose nametags read like a who's who of the agrochemical industry, enzyme scientist Doug Sammons, a senior fellow at Monsanto who looks like he'd be more at home on the farm than in the lab, talked about biologicals. To a roomful of pesticide junkies (and a few others devoted to alternatives), Sammons said that Monsanto's aim was to not only double crop yields over the next fifteen years, but to also reduce chemical inputs by an ambitious one-third.[5] Small start-up companies harnessing these natural defenses are now being snatched up by agrochemical giants (which is either a good thing or a bad thing, depending on whom you trust). Some fear these natural products might get lost in the shuffle or, worse, never make it to the market at all—killed off by an industry already devoted to their own best sellers. Others say it's the only way to get these products through the costly process required for registration. In either case, no matter what one may think of those who brought us the Green Revolution, buy-in by these agro-giants will be instrumental in this even greener revolution. It would be like fossil-fuel companies becoming green-energy acolytes; if big agribusinesses (known collectively as "Big Ag") point their research power toward greener agriculture, we will get there faster.

We are learning to use nature's chemistry to our advantage, from microbial chemicals that rally the troops, nourish plants, and bolster immunity, to antimicrobials and other chemical signals that push away soil-dwelling invaders. But microbes aren't the only option for new chemicals or new ways to boost agriculture, nor are they unique in their response to chemical signals. Insects and plants, too, retreat, attack, and mate when provoked by chemical cues—volatile molecules that swirl on currents of air like creamer in a coffee cup or drift skyward as a "come-hither" chemical plume. Some insect species that are particularly sensitive to scent can detect these chemicals at exceedingly small amounts—parts per *quadrillion*, or one part scent molecule to one thousand million million molecules. We humans are comparatively bad smellers, able to detect at best parts-per-trillion—and more often we are limited to parts-per-billion.[6] When these molecular messages lace the air, they are picked up by insect antennae in much the way that cellular signals are sent over the airwaves from phone to tower. Some insects can trace a chemical plume for miles. While one chemical might attract beetles, another could cause aphids to scatter, or help moths find their mates. These are powerful cues that, like microbial chemicals, offer alternatives to traditional pesticides. We just need to learn how to effectively decode and deploy them. Rather than spraying broad-spectrum killers (which includes both traditional pesticides and even some microbial pesticides), growers can protect their almonds or wheat or apples by hacking the insect communication system and leading pests astray. Moths, a common insect pest, are particularly good at picking up a scent.

An Attractive Proposition

Naturalist Jean Henri Fabre experienced the magic of these invisible chemical signals on a May morning sometime in the late 1800s when a large female peacock moth emerged from a cocoon he'd collected. Within hours Fabre, who had cloistered the moth under a wire bell-jar, witnessed an enchanting scene. Like a butterfly garden, Fabre's home had been turned into a fluttering frenzy of large, randy peacock moths. "With a soft flick-flack the great Moths fly around . . . they descend on our shoulders, clinging to our clothes, grazing our faces. . . . Coming from every direction I know not how, here are forty lovers eager to pay their respects to the marriageable bride born that morning."[7] It would be several decades before advances in analytical chemistry and other innovations enabled chemists to co-opt the insect's chemical language for our collective benefit. One of the more bizarre innovations is the electroantennogram, which is just as it sounds. Electrodes are inserted into the moth's delicate, feathery antennae so that scientists can read the insect's brain's response to chemical cues.

Fabre's moths were responding to chemicals known as sex pheromones—lures that some insects simply can't resist. Of all the insects known to be drawn to their mates by a humanly imperceptible molecular trail, moths are most sensitive. Their elaborate antennae, protruding like antlers on a ten-point buck, have evolved to detect unimaginably small concentrations of chemicals. For moths attracted to our fruit trees, nut trees, and other crops, pheromones are a sensual Achilles heel—good news for those who grow apples, a local fall favorite, because apples are plagued by moths. And no one wants to find a wormy moth larvae in their apple.

Misguided Mates

Over the past few decades, apples, like so many other fruits, have become almost impossibly blemish free. Who hasn't picked up an apple at the grocery store or farmer's market, spun it around, discovered a scab or spot, and surreptitiously put it back? We demand perfection, while at the same time we demand fewer pesticides. That's a tall order for orchardists, because we aren't the only ones who love apples. From the plum curculio—a mottled little weevil whose grubby larvae burrow into the fruit—to the codling moth, apple maggot, and oriental fruit moth, apples are under attack. These are just the pests; there are also pathogens like apple scab and fire blight. And so orchardists, especially in the Northeast, use chemicals to ensure that our Macouns, McIntoshes, and Cortlands are immaculate. Many of these pesticides kill not only the pests, but also beneficial insects, including helpful predators. Sating our demand for out-of-season fruits, growers may also use post-harvest sprays to prevent everything from mildews to discoloration. Even organic growers, when pressed, will dip into the chemical trough, using sulfur, pyrethrins, and copper. (Approved for organic use, the latter two chemicals in particular, though not highly toxic, are not entirely nontoxic.) Though a good deal of chemical residue is long gone by the time we take our first bite, residues of several pesticides remain. In 2016, when strawberries climbed to the top of the Environmental Working Group's "Dirty Dozen" list, they knocked the longtime pesticide-residue champion, the apple, into the number-two spot. In 2013, the USDA reported that 98 percent of apples tested were positive for pesticide, albeit mostly in very small amounts well below concentrations deemed "acceptable."[8]

"It's the nature of the plant realm," says Jon Clements, University of Massachusetts Extension orchardist.[9] Clements, whose e-mail signature is "aka Mr. Honeycrisp," works with growers to solve pest problems. "We'd all love to grow stuff we didn't have to spray. I wish we could figure out how to do apples organically, but it's very hard. You can beat your head against the wall." And you *can* find organic apples grown on the East Coast, says Clements, but it's the rare orchard that succeeds. Most of the organic apples we find in Whole Foods or the supermarket hail from the West Coast or farther afield, from South America—that is, from climates where they don't have to deal with the number of pests and pathogens that populate the moist East Coast. This summer, even my organic CSA compromised when it came to apples. "You just can't do it here," reiterated a grower who works at the local farm co-op and who had pointed me toward the Serenade. "At least not for fresh fruit. Some places can do organic cider." So what's an orchardist who wants to cut back on pesticides to do?

One option is pheromones. "I would say, if you follow the rules it definitely reduces the chemistry used," says Clements. With application rates of milligrams per acre—less than a pinch of salt (versus pounds per acre for many conventional pesticides)—pheromones are powerful stuff. Both highly specific and nontoxic, they are a good option for growers besieged by moths. Flooding an orchard with the arousing scent of female moths confuses males intent on following a plume toward their intended—because the scent is everywhere. If there is no moth sex, there are no eggs or wormy larvae hatching out on a maturing apple or peach or almond. Scents specific to particular species can be captured and reproduced by chemists, synthesized in the laboratory, and packaged up for release. One of the more familiar large-scale uses of pheromones for pest control is for gypsy moths, an invasive pest whose voracious larvae have been chewing their way across the country, denuding acres of forests, tree-lined suburban streets, and city parks, ever since its inadvertent release from a Boston suburb in the mid to late 1800s. I can recall a few New England summers when the moths were particularly randy; their bristling, spotted caterpillar offspring had rendered stands of oaks and maples looking like an apocalyptic fall. In 2016, when the moths seemed to be having an exceptionally good year in the Midwestern United States, municipalities responded by dropping pheromone-infused flakes from planes and helicopters, hoping to save their trees. In Iowa at least, where small pockets of gypsy moths had begun settling in, the treatments were successful enough that no further pheromone airdrops were required that year. Pheromones have been part of the arsenal to fight gypsy moths for decades (pheromone traps are also used to monitor infestation). Increasingly, orchardists are turning to pheromones to save their fruits.

Apple growers had a pesticide problem well before EWG's "Dirty Dozen" campaign began singling out contaminated fruits. Target insects were becoming pesticide-resistant, and the loss of efficacy required growers to spray more. Meanwhile, a new food-protection law began basing regulation on the *combined* effects of pesticides rather than treating pesticides as if they were applied in isolation of one other. The new regulations focused on fruits most likely to wind up in baby-food jars or in a grade-schooler's lunch bag. This put fruit such as apples in the spotlight. Though pheromones didn't promise to solve all of the pest and associated pesticide problems, they could certainly provide some relief from at least one important pest, the codling moth.

Like the gypsy moth and others, codling moths have a weakness for a sexy scent. So, in the early 1990s, growers, researchers, and entomologists put pheromones to the test. As the synthetic scents were dispersed across orchards in Washington State, Oregon, and California, male moths didn't stand a chance.[10] Though results varied (efficacy depends on both the degree of infestation and the potential for mated moths to migrate in from untreated orchard boundaries) and some orchards used pheromones in combination with conventional pesticides, the trials were encouraging, with some cutting back on broad-spectrum insecticides by up to 75 percent. Pheromones could reduce both pesticide application and insect damage. Two decades later, codling moth pheromones are becoming an increasingly popular treatment. In 2015, a government-funded pilot project in Quebec, Canada, alerted apple growers to the benefits of depriving moths of their mates. One grower who says he used to spray against codling moth "five, even six, times a year" managed to cut back to a single spray. "This year," he told a reporter, "we hope to reduce it to zero."[11]

California almond growers, too, are finding solace in unsatisfied moths. Nut trees—almonds, pistachios, and walnuts—are grown in orchards that collectively cover more acres than the state of Rhode Island

and represent a multibillion-dollar industry. Navel orangeworm moths are a major pest of California almonds. After females lay eggs on the outside of an almond, the newly hatched larvae drill into the nutmeat. Their almond tailings, bits of ground-up nut, along with split shells are a tell-tale sign of infestation. Larva-infested nuts are bad enough, but it gets worse. Splits provide an opening for *Aspergillus flavus*, a mold capable of producing aflatoxin B1, one of the most potent naturally occurring carcinogens known. When I was in graduate school and we worked with all sorts of nasty chemicals, a note left on the board by one student read, "I'll do *anything* but AFB1." A few parts per billion of the carcinogen in a shipment is enough to require the destruction of hundreds of thousands of tons of nuts. (Grains can be similarly affected.) That's a real financial hit for growers. So, when one of the industry's most effective pesticides for moths began to fail, growers needed options. One alternative was mating disruption. It was working for apples, but could it help almond growers as well?

In 2002, Brad Higbee, an entomologist who had helped the apple industry disrupt the mating of codling moths, was called in to help almond growers. "After twenty-four years at USDA-ARS, I was hired because Paramount Farming [the largest almond grower in California and perhaps the world] was interested in controlling the navel orangeworm moth."[12] Higbee was summoned at a time when growers were not only facing pesticide resistance among targeted pests, but also a market that demanded increasing perfection. A bit over a decade later, the navel orangeworm pheromone was being released by "puffers" (timed, metered aerosol devices), and was wafting over 60,000 acres of almond *and* pistachio trees. Some orchards were able to eliminate orangeworm pesticide treatment altogether, while others combined pheromones with more-traditional pesticide treatments. In 2015, the number of pheromone-treated acres was around 120,000.[13] It was a small fraction of California's orchards, but expectations are that pheromone treatment will grow. Noting the bins of organic almonds at the local Trader Joe's and wondering how those growers handled pests like navel orangeworm, I asked Higbee if this particular mating disruption system was approved for organic growers, too. But the puffers are formulated with carriers and propellants prohibited on organic orchards, which means, says Higbee, "that organic almond growers don't have many options other than orchard sanitation, the foundation for *any* good navel orangeworm program." There are other biologicals, like spinosads and pyrethrum, but Higbee says neither is very effective. (Rather than depending on propellants, apple growers can use passive devices that release pheromones in response to temperatures; such devices are typically accepted for organic use.)

Pheromones can also lure pests to a deadly end, enticing them into toxic traps, or they can alert growers to an impending invasion. Hundreds of thousands of lures across the country provide a heads-up for gypsy moths. For growers using Integrated Pest Management strategies (a combination of natural and pesticide controls), pheromones mean that rather than spraying the whole orchard with insecticides, they might spray only the border or corner where lures have been set, or only when their fruits are under direct attack. Each year, some 20 million traps now lure boll weevils, grapevine moths, bark beetles, stem borers, and dozens of other insect pests to their deaths. Closer to home, millions of pheromone traps keep houseflies, cockroaches, pantry moths, and, perhaps soon, bed bugs at bay.[14] These are powerful, nontoxic pest controls that can knock back populations, reveal their whereabouts, or lure them to their doom. But there's a catch. Even the best method of pest control is useless if it isn't available to growers. For some growers, the application and expense is simply too much, or for organic growers, the formulation isn't kosher, or there isn't yet an effective pheromone on the market. "Isolating and identifying sex pheromones," says Higbee, "may be relatively easy or difficult. It's highly variable and just depends on the species and the complexity of the components." He notes that it's taken some twenty years to identify some of the minor chemicals required to lure male navel orangeworm moths into monitoring traps.

"Pheromone markets are niche compared to insecticides. One develops a product for a single insect for a single crop at a time," says Cam Oehlschlager, a chemical ecologist and pheromone pioneer. Oehlschlager is now vice president of ChemTica Internacional, a company that develops and sells pheromones to growers worldwide.[15] Since pesticide registration is required in all countries, there's a high start-up cost for products despite the relatively small markets. Even when a new product is clearly effective, "not many significant distributors want to invest in distribution for small markets," says Oehlschlager. The lack of investment in products that are both effective and specific can be a downside on the business end. Even so, pheromone use is growing slowly, by 5–6 percent (average annual growth), but according to one of the last available reports, pheromones have captured just 2–3 percent of the insecticides market (not all that different from twenty years ago). The use of pheromones on orchards is promising, but, as with other biologicals, there is plenty of room for growth and an abundance of products waiting for their day in the field. Oehlschlager says that some 7,000 pheromones and attractants have been discovered for over 2,000 insect species. Hope is in the air—if it can be captured, decoded, reproduced, registered, and marketed.

Mixed Signals

Meanwhile, on farms across Africa, growers have been benefiting from an even more peculiar chemical conversation. Rather than using *intra*-species sexual trickery, or even pitting one insect species against another, British scientists are crossing signals from plant to insect.

John Pickett is a chemical ecologist and scientific leader at Rothamsted Research in Harpenden, England. While I sit with recorder in hand, Pickett nearly hovers in his seat. He is a guy ready to spring into action. Several times, as we chat in his office, he does just that—jumping up to aid a colleague or grab a recording of his jazz group when I notice that he plays trumpet. "You must know of Avogadro's number," he says.[16] I do, sort of, and so I shrug and shake my head. It is a number that all scientists learn, but is really more of a concept. Avogadro's number describes, for example, the quantity of molecules in eighteen grams (a little over half an ounce) of water, which turns out to be a lot. The point is to impress upon us just how small and numerous molecules are. "Avogadro's is 6.02×10^{23}. An enormous number. An incomprehensible number . . . and animals like these aphids can detect molecules right at the bottom end of this—they can detect *molecules*," says Pickett, marveling at the economy.

Pickett has spent a career studying the chemical language of plants and insects. In 2008, he shared one of the world's top agricultural awards (Israel's Wolf Prize) with two Americans. One of his top achievements is a method called "push-pull": playing plants against insects, causing chemical communication havoc. Despite Pickett's enthusiasm for these so-called semiochemicals, though, he isn't imagining a pesticide-free future any time soon: "Where would we be without pesticides? Many more people starving . . . losing at least 30 percent of food to pests, diseases, and weeds."[17] But he is acutely aware of the benefits of allying with nature. Why spend hundreds of thousands of dollars and countless hours to find that one new pesticide, posits Pickett, when you can query nature? "That," he says, "is my *raison d'être*." Push-pull puts the power of plant and insect chemical communication on display.

In sub-Saharan Africa, where small-scale growers (mostly women) struggle to provide food and income for their families, push-pull is touching off a *really* green revolution. For the past few decades, Pickett has been working with his colleague Zeyaur Khan and farmers in Kenya, Tanzania, and other East African countries to pit plant against insect.[18] The results seem almost too good to be true. On fields of maize and sorghum, both staple crops, the stem borer moth is particularly problematic. Like other moth offspring, its voracious larvae feed on young leaves and tunnel into stems. Affected plants are stunted and harvests are reduced for farmers already living on the edge. If stem borers weren't bad enough, crops were also infested with African witch weed, or striga, an exceedingly prolific parasite that feeds upon plant roots and saps crops of their strength. In the mid-1990s, representatives from the Gatsby Foundation (a charitable organization devoted to plant science and to Africa, among several other causes) came to Pickett: "They said, 'we'd like to get some of your elite science into Africa to get sustainable food production and alleviate poverty.'" If successful, the increased harvest would mean more food for tens of millions.[19]

Farmers in the region were already intercropping, says Pickett, harvesting bean crops between maize crops, so it was a good place to begin experimenting with other potentially beneficial crops like molasses grass. "When plants are damaged, they send out an SOS which warns incoming herbivores [plant-eating insects] not to go there because they are already colonized; and it also attracts parasitic wasps." Molasses grass does this, pushing pests away and pulling in beneficial predatory wasps, a sort of twofer. Rather than sacrificing the food crop, the grass takes the hit. "We would make the crop, which was largely intact, *seem* like it was under attack." And there was more. Along the borders, Napier grass and Sudan grass *attracted* the pests while incidentally providing forage for cattle and goats. "It worked brilliantly well. Then purely by chance we found out one of the legumes that we used [for nitrogen fixation] was amazingly good at controlling striga, so we captured all these constraints in one go." Insect and weed control, nitrogen-fixing bacteria, *and* forage—all without the addition of synthetic chemicals.

Pickett and colleagues want to bring this natural technology, already adopted by over 100,000 farmers, to millions. Anticipating a changing climate, growers are already experimenting with more drought-tolerant plants that are better able to thrive on degraded land. There's only one caveat: these are small-scale African farms. "Iowa, Missouri. You can't do it there." The methods aren't amenable to industrial farms supplying McDonald's and Kmart, or even the local supermarket. "So," says Pickett, "we've got to deliver it by genetic modification." The experiments—which attracted plenty of controversy in GMO-shy England—are already in motion.

Reality Check

We are engaged in a war over food: it will be consumed either by pests or by us. As Pickett points out, the ever-growing number of mouths to feed means that we are pushing agriculture harder than ever. Agriculture is the largest land-user on Earth, utilizing 38 percent of our planet's "ice-free" terrestrial surface (crops cover about 12 percent, while pastures cover about 26 percent), and yet too many of us still don't have enough to eat.[20] As our population increases, many more will be without adequate food to eat or water to drink. One report suggests that future demands will require a doubling of food production, while others suggest that the problem is a more complicated matter of how food is grown and distributed.[21] But whether on large farms or small, here in the United States or in Europe or Africa, growers are engaged in an ongoing battle with blights, insects, and weeds (not to mention the changing climate). For nearly a century, farmers have been running on a chemical treadmill—and it is time to break the cycle.

No matter who or where they are, farmers win a hard-fought victory every time they bring a crop to market. Tens of thousands of different *kinds* of pathogens, insects, and weeds dine upon, lay eggs on, and steal nutrients from our food crops. By contrast, humans are at the mercy of a mere 1,400 known pathogens. But only when supermarket prices skyrocket as orange groves are leveled by disease, or blight takes the tomatoes, or we are greeted by maggot-like corn borers, do we take note of the trials facing our farmers. Working on the front lines of the world's food supply, today's growers face a daunting task. They have inherited the system of agriculture created by the Green Revolution—a system that can feed billions, but also requires synthetic fertilizers and encourages large-scale monocrops and reduced rotation, turning crops and orchards into pest-magnets. Growers fight back with millions of tons of fungicides, bactericides, insecticides, and herbicides. But despite all the chemicals, pests and pathogens still cause hundreds of billions of dollars in crop losses here at home and around the globe. Meanwhile, trade and travel introduce new pests, and insects, fungi, and weeds treated one season after another with conventional pesticides evolve resistance—requiring even more pesticide. As with antibiotics, a number of go-to agricultural chemicals are losing their efficacy. Resistance, combined with consumer demand for food with little or no pesticide residue, is a powerful reason to seek out natural remedies. For nearly a century, farmers have been running on a chemical treadmill and it is time to break the cycle.

Agriculture is currently undergoing a second, *greener* revolution. This isn't about organic versus conventional agriculture. It is about growing healthy, affordable food while reducing human and wildlife exposure to potentially harmful chemicals. Toxic pesticides are not the only option; nature produces plenty of her own killers. From semiochemicals to predatory insects and bacteria, these natural allies can move us toward a safer, more prosperous future. As encouraging as biologicals are, McSpadden Gardner adds one cautionary note: none of these products is a cure-all. Natural defenses are simply one part of developing an environmentally and economically sustainable food system. But there is plenty of innovation and hope. The few strategies I describe here are only a fraction of those that are making their way from laboratory to the farm. Some will succeed. Others won't, but they will all add to our collective knowledge of how to grow food more effectively with less impact on the planet. “The book isn’t finished yet,” says Clements. He’s right. And one of the most controversial innovations in agriculture may be one of our best bets for reducing pesticides: genetic modification. New techniques that may someday help put the GMO debate to rest are fast becoming a reality. This politically divisive and endlessly fascinating topic is explored in the next chapter.

CHAPTER 5

Provocation: Disease-Resistant GMOs

Ryan Voiland has been growing and selling tomatoes since middle school, setting up a roadside stand outside his parents' home. A decade or so later Voiland—a soft-spoken, thirty-something organic farmer with a degree from Cornell—was an award-winning tomato grower. "That first year was remarkable," recalled Voiland, cracking a rare smile. "We heard about the Massachusetts Tomato Contest . . . had a good crop and managed to send in some specimens."[1] Red Fire's organic tomatoes won five out of twelve awards, more than any farm, organic or conventional, had ever won in a single year. Red Fire, now a successful Community Supported Agriculture farm (CSA), grows more than 150 different tomato varieties, offering them up for tasting at their annual Tomato Festival. But in 2014, a fungus-like disease called late blight that had been jumping from one farm to another eventually hit Red Fire. Tomato crops died within days. Rows of once-lush plants resembled vegetative versions of zombie armies: upright stalks studded with browned, blight-infested leaves. Large brown spots bloomed on the fruits, turning them soft and unsellable.

The 2014 outbreak left local tomato fields in tatters, but it wasn't the worst case of the blight to hit Red Fire. In 2009, late blight caused massive crop loss. Red Fire took a financial hit, but Voiland had plenty of company that season as the blight ripped into tomato plants all along the East Coast. Chef and author Dan Barber penned a *New York Times* op-ed about the outbreak, "You Say Tomato, I say Agricultural Disaster." The article was just one of hundreds published that year. "I, myself," wrote Martha Stewart in a 2009 blog, "have lost 70 percent of the fifty different varieties in my garden. . . . Many of the beautiful heirloom varieties, which were planted, never had a chance."[2] Stewart's post was accompanied by an image of an ugly, diseased tomato, a far cry from the doyenne's trademark perfection. Late blight, caused by the fungus-like *Phytophthora infestans*, stands apart from the typical specs, spots, and cankers that plague tomatoes. Its widespread distribution is a cautionary tale of big-box stores, mega-plant nurseries, and the delivery of plants potentially infected with a recalcitrant pathogen, which introduced a new problem for tomato growers in the Northeast.[3] Ever since, growers can almost expect blight at some point in the season. The disease has made it difficult for both organic and conventional growers to bring this summer favorite to market. But blight isn't just a threat to tomatoes, a crop we could arguably do without. The pathogen also attacks potatoes: one of the most important food crops in the world.

That 2009 outbreak may have been the first to hit Northeastern tomatoes, but it wasn't the first time *Phytophthora* went pandemic. Nor were tomatoes the first vegetable (or fruit) to be taken by blight. Over a century ago, a mysterious potato disease spread across Europe like wildfire. Healthy plants died within days. Potatoes in the ground turned putrid. Massive crop losses led to widespread starvation, adding to the Grim Reaper's toll from typhoid, typhus, cholera, and other diseases. Some one million Irish died and more than a million sailed for distant shores. Blight had made its way to Europe (by way of the New World, where potatoes, like tomatoes, originated), touching off the infamous Irish Potato Famine that lasted from 1845 to 1852. Since its emergence on potato fields, blight has remained the bane of farmers around the globe. While growers like Voiland might suffer for the lack of summer tomatoes, the population at large will not. Potatoes are another story. *Phytophthora* has become one of the most destructive agricultural pathogens on Earth, causing havoc to growers from the United States to Europe and Africa.

Each year, blight robs food from 80 to 100 million individuals globally.[4] Translated into dollars, late blight is a nearly $7-billion disease. William Fry, a plant pathologist at Cornell University, has been tracking the pathogen for decades. He's a gregarious guy who seems to love his work. When I ask how long he's been at it, Fry laughs and says, "Oh, that's embarrassing," but, as he approaches his golden anniversary studying blight, his contributions to the field are impressive.[5] Fry knows about the grower's struggle with blight. However careful and hygienic growers might be, says Fry, they will eventually have to turn to fungicide. Depending on conditions, growers make anywhere from six to eighteen applications of fungicide a year. Of the big three crops—wheat, corn, and potatoes—potatoes are the most chemical-intensive. A decade ago, approximately 2,000 tons of fungicide were applied to potatoes in the United States alone. Even organic growers apply copper—the "natural" fungicide that is toxic not only to blight but also to the helpful soil microbiota. "As long as blight is around with the given cultivars that we have," says Fry, "we have to use fungicide. . . . People have been trying to make a resistant plant for over a century." Still, the breeding process can take decades, and resistance is too often fleeting. Eventually blight evolves its way around the plant's resistance. "This has happened many, many times in potato-breeding programs around the world . . . to the point that the use of those resistance genes has contributed essentially nothing to late blight suppression." It is almost as if late blight evolved to evolve. The organism is incredibly prolific, providing ample opportunity for mutation and adaptation.

Even in a field where infection isn't yet visible, says plant pathologist Jack Vossen, blight may be sending out hundreds of thousands of spores. "If you look at the genome sequence, it has a core genome," says Vossen, "and a rapidly evolving peripheral genome where a lot of recombination takes place."[6] It is very likely that, within a cohort of spores released from an infested potato field, a few spores will contain the genetic goods to resist a potato's own best resistance. Late blight is an uber-pathogen that twists and flexes its way out of danger, like a genetic gymnast. And even when new blight-resistant potato or tomato varieties become available, both Vossen and Voiland say, they aren't always a hit with farmers, processors, and consumers, who tend to prefer heirlooms.

Fortunately, there is a way to develop late blight–resistant potatoes in cultivars that are popular with farmers, processors, and consumers alike. This brings us to one of the most contentious issues in agriculture today: genetic modification. An astounding diversity of genetic traits exists in plants, animals, and microbes. Within life's collective genome are genes for life-saving drugs, essential nutrients, and antibiotics. If harnessed safely and ethically, there are genes that may also provide resistance to pest and pathogen. Some of these traits, for example, can reduce the amount of chemicals sprayed onto our fruits and vegetables.[7] Yet there are plenty of opponents who are certain that genetic engineering is like selling the soul of a plant to the devil. When I ask friends, scientists, and nonscientists alike about their feelings toward genetically modified organisms (GMO), the responses are mixed. "Seems like a good idea, but are they really safe?" "I don't trust anything Monsanto does." "I don't know, what do *you* think?" But GMOs are not one thing. Introducing pathogen-resistance genes can reduce pesticide use, while introducing herbicide-resistant genes can *increase* the use of a specific herbicide (as with Monsanto's Roundup Ready). Engineering is used to grow more-nutritious food but also food that simply grows faster (like GMO salmon).

It is hard for the consumer to feel comfortable with a technology when even scientists are still bandying about the pros and cons. Some worry about the sanctity of DNA, or the potential for pollen to contaminate non-GMO crops, or the safety of ingesting corn and other products that can produce Bt toxin. Others argue that it is more accurate and safer than some traditional breeding practices. The Union of Concerned Scientists, an organization that describes its mission as "strengthen[ing] American democracy by restoring the essential role of science and evidence-based decision making," urges caution.[8] But their messages are mixed. They point out that health risks are often exaggerated or presented in alarmist terms, writing: "There is no evidence, for instance, that refined products derived from GMO crops, such as starch, sugar, and oils, are different than those derived from conventionally bred crops." But, they also point to unintended consequences: a once-controllable weed turning into a superweed, or the insertion of a gene producing an allergenic protein. (For example, a soybean intended for animal feed was nutritionally enhanced with a gene from a Brazil nut, but the product never made it to market.[9]) "For one thing," they write, "not enough is known: research on the effects of specific genes has been limited—and tightly controlled by the industry." Their main concern is the engineering of herbicide-resistant crops, particularly those resistant to Roundup. As farmers used the herbicide on an increasing number of resistant GMO crops, weeds eventually evolved resistance. In response, growers increased application rates or switched to more-toxic herbicides. These herbicide-resistant crops are one GMO product that many, even some supporters of GMO, agree is problematic.

Meanwhile, in 2016 more than 100 Nobel Prize winners went on record urging Greenpeace, one of the more vocal GMO opponents, to end its opposition to genetic modification. The laureates, a collection of chemists, economists, biochemists, and other medical researchers, stress the precision of *engineering*, compared with *breeding*. The group is particularly supportive of golden rice: rice engineered to produce a precursor of vitamin A. They don't mince words:

> Scientific and regulatory agencies around the world have repeatedly and consistently found crops and foods improved through biotechnology to be as safe as, if not safer than, those derived from any other method of production. There has never been a single confirmed case of a negative health outcome for humans or animals from their consumption. Their environmental impacts have been shown repeatedly to be less damaging to the environment, and a boon to global biodiversity. . . . How many poor people in the world must die before we consider this a "crime against humanity"?[10]

Greenpeace countered with a claim that the Nobel laureates' poster plant, golden rice, was a failure, a venture that they say could exacerbate malnutrition by encouraging rice-only diets. "The only guaranteed solution to fix malnutrition is a diverse healthy diet. Providing people with real food based on ecological agriculture not only addresses malnutrition, but is also a scalable solution to adapt to climate change."[11]

There are scientists, growers, and others who are frustrated that, after nearly four decades, we are still debating the pros and cons of GMOs despite a preponderance of scientific studies concluding that engineered crops are safe for consumption. These advocates are convinced that engineering is the only way to feed 9 billion humans and still have some space on the planet to make a home, to play, or to let wildlife be wild.[12] There is enough about GMOs to fill thousands of books: pro, con, and everything in between. Something for everybody. My goal is to explore pesticide alternatives. How do we reduce chemical residues on our foods and in the soils? How do we reduce the resistance that sends growers into a vicious circle of chemical application? For many crops susceptible to aggressive diseases like blight, genetic engineering is currently one of the only effective ways to do this. And the potatoes engineered by Jack Vossen and colleagues may be a product that even GMO opponents can swallow.

Gene Swap

We are all genetically modified. We owe our lives to countless mutations popping up in genomes over billions of years. Mutations are natural: mistakes that happen each time DNA replicates. Most are repaired by a busy network of enzymes that comb a replicating genome for errors and fix them. A handful are not corrected. Those are the mutations that, should they occur in sperm or eggs, may pass from one generation to the next. A human zygote—after sperm and egg have combined—has on average 130 mutated DNA base-pairs out of more than 6 billion.[13] Many are neutral, neither good nor bad, and are of little consequence in the process of natural selection. Some are lethal, killing off the developing zygote, embryo, plant, or animal before it has a chance to reproduce and pass those genes on to the next generation. A precious few are beneficial. Perhaps an animal tolerates a new food, or a plant sends out more seeds, or grows faster, or resists disease. For eons, plant breeders have relied on mutations bearing these genetic gifts. Corn, tomatoes, and carrots would not taste so sweet if not for humanity's tinkering.

Mutations can happen simply by mistake, or they can be caused by outside factors. The sun's ultraviolet light causes mutations. So does aflatoxin, the toxin produced by some aspergillus fungi that grow on wheat, peanuts, and other crops. Ionizing radiation, like the kind released from radon gas, can rip apart DNA as electrons zing through our bodies and into our cells, potentially causing mutations: a risk to those of us whose houses sit atop granite ledges and foundations. This is the kind of radiation that most people feared when the Fukushima Daiichi plant burned. And it is the kind of radiation directed at our fruits and vegetable seeds to create a better product through mutation breeding. Each year, in support of a local school, I buy a crate of organically grown Rio Starr grapefruit. Its deep rose-colored flesh is the product of a mutation induced decades ago when some ancestral seeds were bombarded with ionizing radiation—intentionally. The mid-1900s was the dawn of the atomic age, and exposure to radiation was a powerful way to create countless random mutations. By the late 1950s, Atomic Gardens, concentric circles of target flowers, fruits, and herbs planted around a central source of ionizing radiation like an atomic age fig leaf, represented a peaceful application of a technology that could do terrible things.[14] Those pink grapefruits, one of the better known products of the atomic age, are now sold as both conventional and as 100 percent organic. Half of California's rice crops are products of this so-called mutation breeding (also produced using mutagenic chemicals). But these are not "GMO"—at least, not technically.

Here in the United States, genetically modified organisms are produced, by definition, through the technological process of introducing DNA that has been combined from different sources outside of the cell. And so, those crops created by radiation (or in some cases, potent carcinogenic chemicals) are not GMO, but are considered just an extension of traditional breeding. This is the case for thousands of fruits, vegetables, and cereals. Some breeders say that the method offers an important alternative in a world averse to GMOs. It's a way to generate mutants that resist disease, tolerate harsh environments, bear more seed, or produce a sweet ruby-red fruit.[15] But mutation breeding is scattershot, as random and unpredictable as nature's own—perhaps even more so. Along with the beauties are the monsters. In those concentric rings were plenty of sickly plants, some with induced tumors, and others too weak to grow. According to a review by the National Academy of Sciences, mutation breeding is more likely to cause genetic disruption and unintended consequences than any other method of plant breeding, including genetic engineering. Yet crops produced in this way are unregulated in the United States, meaning that regulators don't ask for any verification that the crops are harmless.

Genetic engineers don't rely on chance mutations. Instead, they seek out desirable genes across life's collective genome. Technically, genetic modification allows for the transfer of desirable traits from a non-crossable donor species (in other words, a match not made in nature) to a target plant or animal. Plant, animal, and microbe are all fair genetic game. For nearly four decades, ever since biochemist Paul Berg first recombined DNA from two different viruses, genetic engineers have been cutting genes from one organism and inserting them into another. From bacteria and virus into plant and animal; from one plant into another; even one animal to another. It's a process akin to cutting words from one document and pasting them into another—through sometimes less precise, depending on the technique used. Genetic words could end up anywhere in the genome. Other words might be interrupted, though these can be selected against. Or genes that are used as a tag, inserted alongside the desired gene (say, for disease resistance) to show that genetic transformation has occurred, might remain.[16] Many GMO technologies also rely on a bacterium, *Agrobacterium tumefaciens*, to transfer genes. This is a bacterium that, in nature, inserts a small fragment of its DNA into a plant's genome. By inserting the desired gene into the bacterium and then encouraging it to infect the target plant, scientists use the bacterium like a living UPS truck with genetic cargo. But in this case, a bit of the transfer vehicle can be left behind. One concern about traditional genetic engineering technology is the remainder of these foreign bits of DNA from *Agrobacterium* or from the antibiotic tags.

Genetically engineered products are all over the map. Among the newest products are potatoes that bruise less easily and produce smaller amounts of acrylamide, a "probable human carcinogen," when fried.[17]

Microbes in particular have been engineered to become living factories that churn out drugs, vaccines, and even rennet, an enzyme once isolated from the stomachs of nursing calves, that is essential for making hard cheese. Some products are, for better or worse, billion-dollar successes, while others die, victims of consumer's choice. The failure of the first GMO vegetable portended a difficult future for GMOs. Calgene's Flavr Savr tomato made its debut in 1994 and was pulled from grocery shelves in 1997—a casualty of either consumer concerns or poor industry management, depending on whom you read. While the tomatoes were proudly sold as products of genetic engineering, its producer seemed to have focused more on engineering the tomato, rather than the actual, difficult business of growing and shipping enough tomatoes for market. Meanwhile, their presence on the grocery shelves prompted the first call for villagers to storm the castle, pitchforks in hand, to protect us all from this "Frankenfood."[18] But then came Bt corn, a product that would raise the stakes and the profile of GMOs.

Chronicling Monsanto's efforts to produce pest-resistant crops, Daniel Charles, author of *Lords of the Harvest*, wrote of agricultural bioengineers who saw themselves as green "revolutionaries."[19] These were scientists who had come of age during the 1970s, when biotechnology could possibly rein in the flood of pesticides decried by Rachel Carson. Many still see this as a solution to our pesticide addiction—if we can get past the rhetoric, fear, and legitimate concerns of multinational ownership and the unknowns of a new technology. Bt corn, with its toxin-producing gene cloned from a strain of *Bacillus thuringiensis*, was among the first *trans*genic (across species) GMO crops available. Growers, including organic farmers, have relied on the little rod-shaped microbe and its larvicidal toxin as a natural pesticide ever since its discovery in the guts of pesky moths. Monsanto's acquisition of a Bt-toxin gene was an opportunity to improve upon nature; rather than relying on repeated spraying, what if the crops themselves carried the toxin? First isolated in the early 1980s, the gene, writes Charles, was the perfect vehicle for a fledgling industry.[20] Inserted into crops like cotton, corn, and potato, the Bt gene provided growers with an alternative to insecticides aimed at specific pests.

If pesticide reduction was the goal of Monsanto's early bioengineers, Bt was a shining example. Developed in 1996, Bt corn now represents 81 percent of the US crop and has reduced pesticide use by over 100 million pounds over a fifteen-year period. Globally, it is credited with reducing pesticide use by nearly 40 percent, while increasing yields by over 20 percent.[21] But there is controversy over the environmental risks of Bt due to the potential for toxicity to nontarget species, the chance of transfer to wild crops, and the evolution of resistance in target species. Even so, Bt corn remains a widely used GMO crop.

There are also plenty of crops engineered to resist herbicide. Over 90 percent of soy crops and nearly 90 percent of corn is Roundup Ready. Crops are bred to resist the herbicide so that growers can kill off weeds even after crops have sprouted. The development of Roundup Ready crops was like a shotgun wedding for the farmers buying into the cropping system, and some of Monsanto's bioengineers, suggests Charles, regarded the project with scorn, seeing it as a potential stain on the environmentally "clean" nature of biotechnology. The crops are the antithesis of those engineered to reduce pesticide use. While their creation has allowed growers to forgo tilling—a good thing for soils—it is also responsible for increased use of the herbicide. And, as Monsanto offered not only Roundup Ready soy, but also Roundup Ready corn, cotton, sugar beets, and alfalfa, the steady pressure placed on weeds by the herbicide contributed to the ensuing resistance problem. Between these two opposing GMO poles, there are over 100 million acres of corn, cotton, canola, soy, and others in nearly thirty countries (the majority of which are developing countries) engineered to resist pests and herbicide, and in some cases both.[22]

Despite the controversy over Bt crops, pesticide reduction remains a compelling target for genetic engineers.[23] What if orange trees resisted citrus greening, caused by an insect-borne bacterium that is threatening Florida's multibillion-dollar orange industry? Growers who might once have had some success with biologicals are now just trying to save their livelihood with the aggressive use of pesticides. This type of all-out warfare can turn groves into ecological wastelands, bereft of potentially helpful insect predators.[24] What if, instead, apple trees could resist scab, a notoriously difficult pathogen for organic growers; or wheat and other cereals could resist rusts and aphids and other parasites making a living on our crops? Genetic engineering may help reduce the casualties, enabling growers to rely more on biological controls rather than broader-spectrum pesticides. Recall chemical ecologist John Pickett's push-pull system described in the previous chapter. What if wheat crops could be engineered to release a pheromone that drove away aphids while attracting predatory wasps? Rather than expressing a toxin, Pickett's crops, which have been field-tested (though not entirely successfully), are armed with insect pheromone.[25] Or, what if engineered crops were indistinguishable from crops produced the old-fashioned way—through breeding? Without a trace of foreign DNA from antibiotic tags or *Agrobacterium*? The technology is more accurate than classical breeding, certainly more so than mutation breeding. Is there a way to make it more palatable to the public?

Super Spuds

There is no question that transgenic crops can reduce the need for pesticides. And yet consumers push back, afraid of hubris gone wrong. We are wary of creating living things that do not exist in nature and of multinational control over something as essential as food. Neither of these are trivial issues, but there are alternatives that may sidestep some of the most basic reservations about GMOs. One is to produce crops by repairing defective genes rather than adding foreign ones.[26] Another is to produce pest-resistant crops that are no different from those that could exist in nature. This is where Jack Vossen and colleagues come in. In 2015, the group field-tested the world's first engineered late blight–resistant potatoes that are, they say, indistinguishable from the all-natural potatoes we love to bake, fry, and mash. These potatoes could have been bred conventionally—except that it would have taken three decades. Vossen's group did it in three years.

Vossen is both proud of the work and understandably frustrated by the potential for pushback by anti-GMO advocates. Genetic modification, he says, was never the end goal. "The goal was late blight resistance in potatoes." Rather than seeking resistant genes from any plant, Vossen and colleagues confined their search to the nightshade (or *Solanum*) genus, which includes potatoes. By using plant relatives that might naturally interbreed, says Vossen, "we are compiling a completely natural product which cannot be distinguished from crossbreeding. Only we can do it ten times faster." And they can do it *without* leaving a trail of foreign genes, including those from *Agrobacterium*, which is what distinguishes Vossen's potatoes from traditional transgenic crops.

This process, termed *cisgenesis* (from "same" and "beginning"), has some distinct advantages. By producing crops that are genetically "cleaner," the process avoids one of traditional GMO's sticking points. Cisgenesis introduces new traits into familiar genetic territory rather than into a foreign landscape, reducing concerns about unintended consequences. Plus, it produces a crop with useful characteristics sooner. Potatoes can be crossed using classic breeding techniques with relatives, many of which carry disease resistance. But breeding with wild family members carries genetic baggage, including undesirable traits like a bitter taste and the presence of toxins. Before such plants can be marketed, the baggage must be bred out. That kind of backcrossing can take decades. With a threat like blight, growers don't have decades. Since its occurrence in tomatoes in 2009, blight has become an annual occurrence along the East Coast; history provides a glimpse of its ability to spread like wildfire. Moreover, late blight's adaptability means it can evolve its way around naturally bred resistances while rendering one fungicide after another useless. Much like influenza, blight is a slippery pathogen to catch and kill. But it is not invincible.

Solanum plants in the wild have managed blight for eons. One reason is that they naturally grow in separate patches of genetically diverse individual plants. In comparison, monocrops change the ecology; growing thousands of genetically identical individuals on each acre is a crop failure waiting to happen. And, just as we have our immune systems, plants too have evolved immunity. In potatoes and their relatives, resistance genes produce receptor proteins that specifically interact with *Phytophthora*, suggesting that the sole activity of these genes is to recognize the pathogen. Once alerted to invasion, the infected cell essentially commits defensive suicide. Because blight survival requires living tissue, induced cell death stymies the pathogen.

As good as the system is, the plant's receptors and the *Phytophthora* proteins are locked in an ongoing evolutionary war. Played out in the wild, like an ever-changing game of cat and mouse, blight evolves different sets of proteins to hide from the plant's immune system while plants hold their own by evolving receptors that once again recognize blight. It's a different game with monocrop potatoes, though. Bred for size and taste and planted anew each year with seed potatoes, our spuds don't stand a chance. With an ever-changing pathogen like blight, the crop's receptors need to be almost as flexible. "This is what we offer as breeders," says Vossen, "a variety of resistance genes." Not only do Vossen and colleagues breed resistance into potatoes in short order, but they can also insert *sets* of resistance genes. Multiple genes make it less likely that blight will escape. (Physicians employ a similar strategy when they treat cancer with combinations of chemicals.) "Simply put, potatoes are not able to get there with crossbreeding."

Images of field tests with the cisgenic potatoes are striking: robust plants filled with emerald leaves next to the zombie soldiers, their crumpled brown leaves still clinging to the stalks. This is blight resistance without the foreign genetic "baggage." Yet the crops retain the stigma of a GMO. "As long as *cisgenesis* is considered genetic modification," says Vossen, "I don't think they'll be sold on the EU market. If they are considered distinct and regulated as distinct from GMO, then yeah, there will be a future." The Wageningen Institute in the Netherlands, where Vossen works, has even taken pains to assuage concerns about a corporate takeover by retaining the intellectual property (IP) at the institute and offering non-exclusive licenses to parties that would like to work with the genes or with resistant plants. Vossen says that most people agree that cisgenic potatoes are different than transgenic crops. Even the European Food Safety Authority (EFSA) agrees. At the request of the European Union, the EFSA Panel on Genetically Modified Organisms was asked to deliver an opinion on cisgenic crops. They did, finding the crops not guilty. The hazards lurking in cisgenic crops, concluded EFSA, are no different from those of conventionally bred crops, while transgenic crops may introduce new, unknowable risks.[27] Despite the vote of confidence, Greenpeace remains stubborn in their rejection of the technology. This frustrates Vossen: "We have the same goal: we both want to reduce pesticides."

Cisgenesis is one way forward. J. R. Simplot's recent product, Innate potatoes, engineered to resist bruising and produce less potentially carcinogenic acrylamide in the fryolator, uses a process similar to Vossen's, except that the potatoes also have genes modified to knock down the expression of undesirable traits. In 2016, Innate potatoes received FDA clearance, although, according to Simplot, they were still awaiting EPA registration.[28] More options are becoming available for tinkering with genes, in ways that are closer to all-natural. One of these is gene editing—a process like gene therapy, whereby malfunctioning genes can be repaired. In particular, so-called CRISPR technology (clustered regularly interspaced short palindromic repeats—pronounced "crisper") has recently made gene editing so much cheaper and easier that some say CRISPR will "democratize" genetic engineering.[29] With CRISPR, the hope is that researchers can more quickly and accurately disrupt or edit genes by using an enzyme system that bacteria have used for eons to copy and snip out bits of invading viruses (bacteriophages). These same enzymes can be used to target plant or animal cells. As promising as cisgenic multi-gene resistant potatoes and other efforts may be, Fry, who has seen plenty over his career and has a healthy respect for pathogens' ability to evolve their way around resistance, adds a cautionary note: "*Phytophthora infestans* can, and certainly has, overcome multiple resistance genes in one host." So he urges growers and developers alike to not put all their eggs in one basket. Instead, they might combine gene modification with strategies for improving the crop's immune response, or use *low* levels of fungicide to help stymie the emergence of virulent strains.

There are no silver bullets: no single methodology will cure or prevent all disease, not in agriculture, nor in medicine. But each may offer a stepping-stone of progress. There are innovations that will reduce our reliance on pesticides and help preserve our valuable antibiotics. Some are just making it to the market. Others are moving through the regulatory process, a costly but necessary step to verify efficacy and safety. For some products, the testing requirements are overly onerous, and for others, testing may not be rigorous enough. Finding the right balance between caution and experimentation will always be a struggle, but the more closely we work with nature, the more amenable nature may be to our interventions.

If the debate is hot over genetic engineering in the food people eat, it enters a whole other realm of ethics when it comes to people themselves. Yet even as we struggle with the morals of genetic engineering *in* humans, scientists are using the technology to tackle diseases from flu to HIV, and now Zika and Ebola as well. In the case of all these pathogens, scientists are engineering vaccines better able to provoke a response from our own immune system.

CHAPTER 6

Provocation: The Next Generation of Vaccines?

My father, just returned from the US Navy, was a mischievous, apple-cheeked twenty-year-old looking forward to his junior year in college when meningitis struck. It was 1946 and the last thing he recalled was brushing his teeth at home in the bathroom. For the next ten days he lay unconscious in a hospital bed, his body fighting off an invisible army of bacterial invaders. Aided by the new miracle drug, penicillin, he survived, but not entirely unscathed. Shortly after recovering, my father was jolted by seizures, his brain permanently damaged by the infection. For the remainder of his life he managed the condition with a combination of powerful antiepileptic drugs (while baffling his doctors by referring to the electronic brainstorms as a "free high").

Meningitis is a catch-all term for swelling of the tissues surrounding the brain and spinal cord. Specific viruses, fungi, and injury can all trigger the potentially fatal condition, but one of the most frightening and lethal causes is bacterial infection. Bacterial meningitis can kill within a day, is often incurable, and may leave survivors with amputated limbs, hearing loss, or brain seizures. My father was relatively lucky. One of the more obstinate causes of meningitis is *Neisseria meningitidis*, a bacterium adept at spreading in places where people are gathering for the first time: freshmen dorms, summer camps, day care, military barracks. Some 5–20 percent of us carry *Neisseria* in our noses and throats and unwittingly spread it to those with whom we share a meal, a drink, or a kiss. Most of us won't get sick. But a few of us may die from the infection, even today.

In the 1960s, annual outbreaks over a four-year period killed fourteen members of the Fort Ord community in California. At the time, the sprawling base was processing upwards of 1,000 new recruits a week—fertile ground for *Neisseria*. One study showed that by the end of basic training—eight weeks of sweating, sleeping, dining, and drinking with their brothers in arms—recruits carrying the bacterium rose from some 20 percent to as high as 90 percent.[1] With no vaccine available there were few options for prevention, save widespread "chemoprophylaxis"—treating thousands proactively with the antimicrobial drug sulfadiazine. These were the days before we knew of the microbiome, or the looming crisis of antibiotic resistance. Inevitably, *Neisseria* strains began resisting the drug and it largely went out of use, though, until recently, widespread prophylactic antibiotics were still sometimes the only option against *Neisseria*.

The ongoing threat of meningitis in military recruits, combined with fears that the disease could spread into surrounding communities, turned *Neisseria* into a target for vaccine makers. The first vaccine, tested on recruits in 1969, was a phenomenal but limited success. It reduced the incidence of disease, but was effective against just one of a dozen known serotypes (members of the same strain, but which are recognized as different by our immune system), half of which cause the majority of disease in humans. Even so, just a couple of years after that first success, meningitis vaccination for US Army trainees became part of the intake routine. A few years later another vaccine became available, this one protective against four serotypes, but still there were limitations. Protection was fleeting in the very young and it didn't provoke so-called immune memory, one of the most effective protective responses; and so it was recommended only for a very limited segment of the population.

My kids were born in the 1990s. By the time they toddled off to school, they had received a slew of vaccines: for measles, tetanus, mumps, polio, chicken pox, and even *Hemophilus influenza* and *Streptococcus pneumonia* (two other important causes of meningitis). But an effective vaccine against *Neisseria meningitidis* had not yet made it onto the recommended vaccine schedule. Then, in 2005, just as they were heading off to the middle-school milieu of new students, sweaty locker rooms, team sports, and shared drinking bottles, a vaccine against a collection of *N. meningitidis* serotypes became available.[2] Though the disease is rare here in the United States, particularly compared with the so-called meningitis belt of sub-Saharan Africa, I felt relieved. One more disease they wouldn't get. Except.

Except for the escape artist, a serotype called meningitis B, or Men B. Though rare, the infection has frustrated vaccine makers for decades. It seems to pop up out of nowhere and it can make someone seriously ill within hours.[3] In 2013, an outbreak at the University of California forced a freshman lacrosse player to undergo amputation of both feet. Four other students were infected, and the university was forced to provide prophylactic antibiotics to 500 students. The very next year, an outbreak that began at Princeton University caused the death of a Drexel University student.[4] In the first months of 2016, Men B hit three different colleges and killed one employee.[5] Even in our "golden age" of disease prevention, Men B has remained intractable—until now. In 2015, the US FDA licensed two new vaccines, which are making inroads against the disease. One of them, a vaccine called Bexsero, could not have been produced without twenty-first-century technology. Advances in genomics are revealing pathogen targets that had once been obscure or even completely unknown, providing new options to combat some old adversaries.

We Are Our Own Best Defense

Vaccines work with our immune system, and the human immune system is extraordinary. It is the product of an age-old battle with pox, syphilis, cholera, and other infectious diseases. Well before there were humans or even prehumans, animals large and small fended off viruses, bacteria, and other minuscule invaders. As a result, one in thirty human genes—roughly 3 percent—encode proteins involved in immunity or defense.[6] When it comes to pathogens, our immune system is like a dystopian society obsessed with detection, destruction, and prevention; busy cells and messengers conduct surveillance and never sleep. It is a system able to sift through the constant bombardment of pathogens and molecules entering our bodies, responding to dangers while largely ignoring the rest. Though our skin, mucus, stomach acids, and many of the microbes we call our microbiome (and the antimicrobials they produce) serve as a first line of defense, these barriers can at times be breached. This is when other "innate" immune defenses—cells that gobble up or kill invaders while releasing an arsenal of chemical defenses—shift into gear. The ongoing struggle may raise our temperature or settle into our joints as we take to bed. This is immunity in action, and it's the price our bodies pay for nonspecific defenses. But this innate response is also buying time for a more specialized response called adaptive immunity.

Within days or weeks of infection, our adaptive immune system begins cranking out specific antibodies that can attack invaders just like predator drones. We also produce killer cells that, like antibodies, target particular intruders by recognizing specific molecules, or antigens. Each cell of this adaptive system is recruited from a huge storehouse of immune cells just waiting to be selected. Deep within our bones is a birthing ground for new blood cells, including immune cells.

When under attack, the cells that produce the most-effective antibodies survive and divide, making clones. The process continues, ratcheting up the specificity and refining the antibody. Within a week or two we are fully armed with the best-fit antibodies. Antibodies circulating in our blood plasma are like a biochemical surveillance system, marking their specific pathogens for destruction. Yet, once a pathogen makes it *into* our cells, antibodies are of little use. Some of the trickiest infections, from HIV to malaria, hide out in our cells, making eradication difficult and disease recurrence almost inevitable. But our immune system is prepared for this scenario as well. Inside our cells is the molecular equivalent of a food processor that breaks down a pathogen's antigen, grabs a bit, and presents it on the cell's surface. The cell is effectively waving a flag: "Hey! I've been invaded, so for the good of the whole I am prepared to sacrifice myself." T-killer cells recognize the display and help the infected cell to self-destruct.

The adaptive system also provides us with immunologic gold: immune memory. Long after the offending pathogen is gone, a set of immune cells continue circulating, acting as sentries. Should the cold that hit us last week come back, we will respond far more rapidly, in days rather than weeks. We may suffer milder or different symptoms than when we first encountered the virus, or we may even remain blissfully unaware of its return. Of course, we are not invincible; sometimes, particularly if the disease is aggressive or we have been weakened, we may lose the battle before the immune system is fully armed. But if we are prepared, we stand a better chance. This is where vaccines come in. Like war games for the immune system, vaccines prepare our immune defensives for future invasions.

Immune Boosters

By the time my kids were six, cells primed by vaccines against at least ten potentially fatal diseases were circulating around their bodies (along with immune cells that remembered who-knows-how-many other viruses and bacteria to which they'd been naturally exposed). Between my parents' generation and my children's, vaccines have prevented hundreds of millions of illnesses and millions of deaths here in the United States. The number of recorded and estimated deaths around the world in the twentieth century is a humbling reminder of our vulnerability: smallpox killed 400 million; measles 97 million; whooping cough 38 million; malaria 194 million; seasonal flu 36 million; and the 1918 "Spanish" flu killed 40 million (at least).[7] Over the same time period, humans have killed just over 100 million of our own in warfare.[8] We cannot vaccinate against inhumanity and violence; but we can against disease. There have been few health interventions so far-reaching across the globe.

Today, roughly 86 percent of children are immunized against polio, diphtheria, tetanus, and pertussis worldwide.[9] But the telltale circular scar of the smallpox vaccination that I wear on my upper arm is foreign to my children. It is the first pathogen and perhaps the only *microbe* that we humans have driven into extinction in the "wild." (Frozen stocks still exist, though there are calls for their destruction.) Nearly twenty-five years after smallpox killed Lillian Barber, a Texas mother of three who was its last victim here in the United States, the smallpox vaccine was dropped from the schedule.[10] A few years later, in 1977, naturally occurring smallpox killed its last known victim in the world.[11] The vaccine was so successful that it brought an end to its own utility. Health workers around the world are hoping to do the same with polio virus and vaccine. These are the success stories. But there are failures, too. Potentially lethal diseases like TB, malaria, HIV, and, until recently, meningitis B have continued to frustrate vaccine makers and leave populations at risk, particularly in developing nations. Others like Ebola and Zika emerge in the blink of an eye and catch us all off guard as we clamor for a new vaccine that we could have used—yesterday. Unfortunately, a decade may pass between the discovery of a new pathogen and the development of a vaccine that is ready for widespread use.

In 2015, when Zika began sweeping across South America before making its way up to the United States, Dr. Anthony Fauci, director of the National Institute of Allergy and Infectious Diseases was asked how quickly we might have a vaccine. Even with an accelerated program and all hands on deck, the best Fauci could offer was maybe eighteen months.[12] And that was because vaccine makers are already partway there, able to build upon an existing vaccine "backbone" developed for related viruses like dengue and yellow fever. With a disease like Zika, which moves across the placenta and into a developing fetus, a year and a half is too late. And had the backbone *not* existed (assuming that this particular vaccine is successful), a Zika vaccine might be years in the making and testing. In June 2016, a Zika vaccine received FDA approval to enter into human trials, the first step in a multi-year process. Other vaccines against Zika will likely follow as they are approved for trials.

The concept of a vaccination is simple enough: vaccines provoke immunity by exposing individuals either to pathogens that have been weakened or killed so that they can no longer cause full-on disease, or to bits of pathogens. But pathogens are wildly diverse, and a vaccine strategy that works for one disease may not work for others. Some are fairly straightforward—for example, injecting weakened or killed polio virus provides lasting protection. (Since 2000, the United States has used only killed polio virus.) When I was vaccinated as a kid, I likely received the next best thing to a natural infection: live but weakened versions of polio, mumps, and measles. A generation later, most of my children's shots were filled with inactivated or killed viruses, or bits of microbes.[13] Kids today do still receive some attenuated (weakened) virus vaccines, notably against mumps, measles, and rubella.

Many of these twentieth-century vaccines began with Maurice Hilleman, a virologist and vaccine developer who spent most of his career at Merck Pharmaceutical. The mumps vaccine my kids received may even be traced back to the 1963 mumps virus that once infected Hilleman's own daughter, Jeryl Lynn. As he tells the story, one night she woke complaining of a sore throat. "Oh my god," said Hilleman, pointing to the glands under his chin and holding out his hands, "her throat was like this."[14] Though rare, a mumps infection can have serious complications, from permanent hearing loss to life-threatening brain swelling. There was no vaccine. So Hilleman raced to the lab and grabbed some swabs. Three years later, he treated his one-year-old daughter Kirsten with a vaccine he had developed from Jeryl Lynn's virus. "Here was a baby being protected by a virus from her sister, and this has been unique in the history of medicine. . . . It was a big human-interest story." Hilleman, who passed away in 2005, is credited not only with developing dozens of vaccines but also saving more lives than any scientist before him. But, as the authors of an article in *Science* about twenty-first-century vaccine development pointed out in 2013, "By the latter part of the twentieth century, most of the vaccines that could be developed by direct mimicry of natural infection with live or killed/inactivated vaccines had been developed."[15] In other words, the most manageable pathogens, like mumps, were under control. What's left for vaccine makers are the problem pathogens. Influenza viruses, for example, evolve so rapidly that one year's vaccine may be ineffective the next. And though vaccine makers have become really good at predicting the upcoming season's flu and recommending which strains to use in the annual vaccine, sometimes they miss. Or consider the malaria parasites that pass through so many different life stages, moving in and out of liver and blood cells evading capture by circulating antibodies. HIV not only strikes at the source by invading and destroying our immune cells, but, like influenza, is also a shape-shifter. And then there is meningitis B, a pathogen that disguises itself in molecules so similar to human that our immune cells see "self" rather than "invader" and back off. These are the challenges for twenty-first-century vaccine makers. They are also confronted with a growing trend of distrust in vaccines. Ironically, vaccination critics are part of a population that has benefited greatly from vaccines, largely avoiding the raft of infectious diseases that plagued earlier generations.

Yet no matter how many lives vaccines save, there is no skirting the issue. Vaccination is a medical intervention. We inject newborns and toddlers—the most vulnerable members of society, who cannot decide for themselves. Some parents worry about their kids receiving too many vaccines at once. Others are concerned by the small amounts of toxic chemicals like formaldehyde and ethyl mercury used to kill or to preserve vaccines. Some believe conspiracy theories about vaccines spreading disease. And many have been frightened by a now-discredited study accusing the MMR vaccine (also developed by Hilleman) of causing autism. Some of these concerns contain an unsettling kernel of truth. A portion of the polio vaccines that my generation—millions of children—received were contaminated with the monkey virus, SV40. Until the 1960s, polio vaccine was grown and isolated from green monkey cells. Hilleman and a colleague discovered the virus; a couple of years later, another researcher showed that the virus caused cancerous tumors in hamsters. By the time vaccine makers had replaced monkey-cell cultures with human cell cultures, an estimated 100 million of us baby boomers had been vaccinated. Fifty years later, despite much suspicion and study, the virus has not yet been shown to cause cancer in humans.[16]

Vaccines made with weakened live viruses tend to induce immune responses that are highly effective and long-lasting. By passing the wild vaccine through nonhuman cells (successively growing the virus in chicken or mouse embryos or in guinea pig cells), the virus evolves over time, becoming better suited to infect its new host rather than a human host. With a vaccine, the altered virus survives and replicates just enough to provoke a response but not to cause disease. Rarely, before a weakened virus is cleared it may mutate back to its original virulent form. By some estimates, one in 2.7 million people who receive the live-virus polio vaccine runs the risk of developing a case of vaccine-derived polio.[17] The live vaccine is no longer administered here in the United States, where the last outbreak occurred in 1979, but it is still considered the most effective vaccine in countries where polio remains problematic.

While there may always be unintended consequences of vaccines, the role they have played (and continue to play) in saving lives over the past century has been huge. Now vaccine makers have the tools to develop increasingly safer vaccines, effective against some of the most obstinate pathogens—and they can do so more rapidly.

Prevention Engineering

Those who devote their lives to making vaccines belong to a vast scientific community including virologists, bacteriologists, protozoologists, immunologists, vaccinologists, geneticists, and others working everywhere from universities to tech start-ups, federal labs to Big Pharma. In just one year, the National Institute of Allergy and Infectious Disease listed over 400 vaccine-related research projects (many being multi-year efforts) supported by a bit over $430 million in funding. And that is only a portion of the ongoing effort to prepare us for an encounter with a potentially deadly or debilitating pathogen. There is research funded by other federal agencies, private organizations, and venture capitalists—not to mention many institutions outside the United States.

Vaccine development is one of our best prospects for protection against disease. It is also a multibillion-dollar business. Currently estimated to be over $20 billion, the vaccine market is projected to rise to over $60 billion by 2020.[18] (In addition to pathogens, vaccine makers are now targeting some cancers—a whole other fascinating topic.) It is a booming but risky venture. Researchers can work for years, with no guarantee that the fruits of their labors will save humanity from Ebola or rotavirus or *C. diff* or malaria or whatever else ails us. These are scientists at the front lines of defense. They are working to discover new vaccines; improve the efficacy of existing vaccines; and make those that work more affordable and available globally. They are immersed in a world where pathogens bump up against immunity, setting off cascades of complex responses: B cells, T cells, antigens, epitopes, and antibodies, as well as myriad biochemical molecules that have evolved as part of an ancient conflict between host and pathogen. Effective vaccines save lives on a global scale. But development can be an expensive slog from discovery to clinical trial—where many promising candidates fail, while a precious few vaccines prevail. Vaccine making is a venture for optimists driven by the potential for doing good on a global scale.

Lenny Moise, director of vaccine development at EpiVax, in Providence, Rhode Island, is one of those optimists. As a self-proclaimed minimalist in a world teeming with antigens, Moise is emphatic that with today's technologies they can do more good with less: "We can't just inject anyone today with something that is not well understood, when we now have tools that enable us to characterize vaccines and the immune response they elicit."[19] EpiVax is one of many small companies (along with academic, federal, and corporate research efforts) devoted to combating pathogens with information packaged into more effective vaccines. They call it gene-to-vaccine or genome-based vaccine development. Some call it reverse-vaccinology, since researchers begin with the *genome* of the pathogen, rather than seeking out immune-active regions on the actual pathogen. With access to informatics and genomics technologies, developers can not only get to know the pathogen down to its genetic sequence, but also predict how our immune cells may respond—well before a drop of vaccine makes it into the syringe. These are next-generation engineered vaccines, ground that was broken by

Maurice Hilleman and colleagues back in 1986 when they engineered a vaccine against hepatitis B based on a *single* viral antigen.[20] Here in the United States, the disease wasn't considered a significant killer, and few, save those whose lives had been touched by it, had heard of it. But the vaccine was big news.

Hepatitis B is passed through blood and bodily fluids and infects billions worldwide. Those at greatest risk have multiple sex partners, use IV drugs, or are health workers. Before a vaccine was available, chronic Hep B was thought to cause 80 percent of liver cancers worldwide. The first vaccine developed by Hilleman and colleagues, licensed in the 1980s, relied on a single antigen (called the Australia antigen) that circulated in the blood of infected individuals. It was the viral version of the Achilles heel: a bit of virus that provoked immunity without the risk of causing disease. The vaccine was a breakthrough. So too was the first source of antigen: infected human blood plasma. Despite rigorous processing that ensured that the vaccine wouldn't be contaminated with any live, whole virus, the "subunit" vaccine was a tough sell in the 1980s when HIV was just emerging as a lethal blood-borne infection that often co-occurred with Hep B.[21] Seeking a more acceptable vaccine based on the same effective antigen, Hilleman collaborated with colleagues at the University of California–San Francisco who not only sequenced the antigen but successfully inserted the gene into yeast cells. The result was a pure source of antigen without the risk of additional blood-borne viruses.[22] Recombivax HV, the first engineered vaccine to be licensed, was released in 1986 and made headline news. It was a new medicine, a harbinger of hope and things to come. Interviewed by the *New York Times*, Frank Young, the FDA commissioner at the time, called the innovation historic: "This development opens the door for the production of other vaccines that have so far been impractical, potentially unsafe, or impossible to make," he said. "Genetic engineers are already working to develop vaccines against such parasitic diseases as malaria, schistosomiasis, and filariasis, and such viral diseases as acquired immune deficiency syndrome," he added.[23] That was thirty years ago. The technology, though promising, wasn't completely there—yet.

Advances in genomics, combined with high-power computing and data-crunching programs, are enabling scientists to comb through the genetic sequences of pathogens, particularly those that have proven elusive to vaccine makers. These new strategies may lead to more-effective and safer vaccines against diseases like malaria, HIV, and TB, and, someday, on-demand vaccines for fast-emerging pandemics like Zika. Some vaccine makers are tinkering with a pathogen's genes, removing virulence genes from whole pathogens like typhoid fever while seeking to prevent microbes from reverting to more pathogenic forms; other vaccine makers, including Moise, are breaking pathogens down to their smallest antigenic bits and building vaccines by stringing together DNA molecules. Moise and colleagues envision a day when vaccines might be even more specific.

Sequencing technology, combined with computer algorithms, could be used to identify and test "in-silico" (computer-modeled) candidate target molecules (called epitopes—the bits of antigen that are recognized by the immune system) against immune sequences, seeking a best-fit match. It's sort of like high-tech, high-speed dating. "As a minimalist," says Moise, "I like that very much. Rather than *more* information than is needed to protect, you can pick out the specific triggers needed—wouldn't that be better?" Not all vaccine researchers agree that this is the best way forward, asking: Why not use the *whole* antigen? They worry that increasing specificity may lead to a pathogen evolving its way around detection. There are also concerns that these methods may not be able to deal with individual and biological complexities, which could send vaccine makers back to the drawing board—at the taxpayer's or investor's expense. In some cases, like influenza, they argue that it would be better to have a vaccine that would be effective against various strains and subtypes: in essence, one vaccine to rule them all (or, at the very least, several of them).

"There are believers and nonbelievers. I see arguments for and against a minimalist approach," says Moise. He cites animal studies that show how minimalism holds promise for treatment of both infectious disease and cancer, and he believes that specificity could also cut down on a vaccine's side effects. As Moise explains, "We can design vaccines today using informatic algorithms . . . to identify the key elements of a pathogen that drive protective immunity."

This sort of hyper-specific approach to vaccine making is having some success in cancer treatment, yet for infectious diseases, Moise says, "we are further away." There is just too much variability among people for this kind of individualized vaccine making on a large scale. Our age, genetics, experiences with the environment, and even our microbiomes shape our immune response, causing each of us to react just a little differently to both pathogens and vaccination. But, adds Moise, "We are [also] seeing what we can do to tweak antigens to make them more effective." Sometimes, even when a vaccine exists, it doesn't work all that well. While most approved vaccines are highly effective, there are vaccines for important diseases like dengue and malaria that are not as foolproof as public health workers would like. For some of these, Moise and his colleagues think they know why.

Our immune system is exceptional at picking out and responding to antigens, but some viruses and bacteria still manage to slip by our defenses. As Anne De Groot, EpiVax cofounder and Moise's colleague, explains: "Viruses like HIV and EBV have built themselves a very successful niche . . . using bits and pieces of the human genome. . . . The 'immune signature' of a virus is often indistinguishable from self. Viruses (and bacteria and parasites for that matter) have found ways to camouflage themselves, to reduce the likelihood that they will be identified, trapped, chewed up, spit out, and eliminated." Pathogens hiding behind antigens that look like us don't make for effective vaccines. But Moise, De Groot, and coworkers believe they know how to make a better vaccine, even against these stealthy pathogens. The majority of vaccines on the market today rely on an antibody response to antigen and this, explains Moise, requires T cells. Some of these pathogens have evolved to take advantage of one kind of T cell in particular (there are several). The T cells critical for antibody-based protection are effector and regulatory cells. Regulatory T cells are tuned in to our own proteins. They help prevent autoimmune reactions, damping down our immune response against self. Effector cells turn up the immune volume; these are the cells that help kick off antibody production. A pathogen displaying an antigen that looks human rather than foreign may cue regulatory T cells to tamp down antibody production. "If we can identify those anti-inflammatory triggers," says Moise, "we can design vaccines using recombinant DNA technology that can eliminate those triggers." The goal is to uncloak the stealth pathogen and trigger a stronger, more effective antibody response.

One of their first targets is the H7N9 influenza. In 2013, H7N9 jumped from birds into people, killing nearly 30 percent of these known to be infected. First identified in mainland China, the human outbreak led to a slaughter of tens of thousands of geese, ducks, and other poultry in an effort to slow the spread of the virus. Should H7N9 evolve a capacity to spread from one human to another, it could go pandemic; this is the kind of scenario that keeps health scientists awake at night. The current vaccine against H7N9 isn't all that effective, and so developing a better version is a high priority.[24] Based on computational tools, De Groot and colleagues discovered that the virus was essentially hiding behind human-like genes. So they stripped H7N9 of its cloak. "We've engineered a vaccine that is more immunogenic [for H7N9]," says Moise, by removing the anti-inflammatory triggers.[25] The vaccine was moving toward clinical trials when we spoke, but all clinical trials require cautious optimism.

Even if this H7N9 vaccine doesn't pan out, there are other benefits of sifting through a pathogen's genome. One is identifying proteins that aren't prone to the kind of rapid evolution common in many pathogens, influenza in particular. A vaccine against this kind of conserved antigen is the brass ring that influenza vaccine researchers have been reaching for. Its discovery might someday mean better control of annual influenza and even protection against future pandemic flu. Rational vaccine design offers hope for managing diseases like HIV, TB, and malaria that have confounded vaccine makers for over a century. While the new approach won't solve all of our vaccine problems, it has already provided some relief for one problem in particular: meningitis B.

Men B evades immunity not by producing human-like antigens, but by wrapping itself in a sugary sheath that is identical to human molecules. Immune cells that recognize this molecule are naturally eliminated or deactivated as a protection against autoimmunity. By sequencing the pathogen, vaccine makers were able to discover antigenic proteins that would otherwise be hidden. Four different antigens found on the majority of circulating Men B would be used for the vaccine (a single pathogen may have several different circulating strains). Developer Mariagrazia Pizza and coworkers reported their groundbreaking findings in the journal *Science*, writing: "In addition to proving the potential of the genomic approach, by identifying highly conserved proteins that induce bactericidal antibodies, we have provided candidates that will be the basis for clinical development of a vaccine against an important pathogen."[26] In 2013, the vaccine was licensed in Europe, though not in the United States. But after meningitis broke out at Princeton and UCSB, the vaccine was offered to students on both campuses. One headline blared: "California students to receive unlicensed meningitis vaccine."

Sold as Bexsero by Novartis, the vaccine (along with another new vaccine called Trumenba) was licensed in the United States in 2015.

Moise is enthused by the opportunities to develop new, effective, and safe vaccines: “We’ll have more successes,” he says. “Many groups have their own approaches to producing vaccines, and they are finding ways to accelerate the processes so there won’t be this lag between emergence of new pathogens and the responses we can make to getting vaccine out to people. The potential for doing good on the global level is amazing.”

Endless Conflict

Novel diseases will continue to emerge as pathogens find new hosts. Some pathogens jump from wildlife that has been forced into our lives as we cut away old forests to make way for cities and agriculture, or simply wood to burn. Others find communities living on the edge, malnourished and vulnerable to disease. Still more are carried around the globe as both travel and population increase. While we must certainly try to prevent new diseases, particularly as we become increasingly aware of our role in creating conducive conditions for epidemics, vaccines offer a measure of protection. They represent a way to better prepare our own immune system for the inevitable run-ins we’ll have with viruses, bacteria, and other pathogens. Effective vaccines are one of our best defenses against existing and emerging pathogens, and there are hundreds under development. Some exist only in code, not yet tested against the complexities of living, breathing biological reality. Others are in clinical trials; if proven successful, they could be providing relief in a year, or two, or three. Technologies like genomics and computational power are opening doors to new strategies. But life is complicated and pathogens are numerous, diverse, and quick to evolve. Vaccine development exemplifies science on the move, poised in the precarious space occupied by the excitement of scientific discovery, mistrust of Big Pharma, and the public’s misunderstanding of the incremental nature of progress.

While we would always wish to avoid illness, whether through effective vaccines or simple good luck, when it strikes, we have to act quickly. The first step is accurately identifying the culprit, which is not as straightforward as it sounds, particularly in a world of emerging diseases. The future of better treatment and prevention, whether in agriculture or medicine, is tied to fast, accurate, and accessible diagnostics—the subject of the last two chapters.

CHAPTER 7

Know Thine Enemy: Images of Disease

SOMEWHERE IN A CARGO HOLD OR CRATE, or clinging to a shoe or leaf, is the cause of the next great plant pandemic. Perhaps it will kill off our tomatoes, or our corn. Maybe it will hit our cherished strawberries or threaten our wheat. On average, roughly 18 percent of the world's food crops are lost every year to pathogens and pests. (Even more is lost to weeds.[1]) Some have been infecting and eating our food for centuries, while others have only recently arrived on our shores. Even with today's technologies for thwarting weeds, insects, and diseases, crop loss is huge (not to mention loss after harvesting). As a typical consumer, I was blissfully unaware of the losing battles waged by growers here in the United States (aside from the regional bout with late blight) until recently. In 2015, a little insect from Asia called the "jumping plant louse" had begun to cause enough havoc to make it into the news cycle. While I grumbled over the rising price of grapefruit (a midwinter favorite), the plant louse was feeding on citrus crops, infecting them with *Candidatus Liberibacter asiaticum*, a bacterium that causes citrus greening, a devastating disease also known as *huanglongbing*. By 2016, 80 percent of Florida's citrus crops had become infected. When the bug and bacterium combo arrived nearly a decade ago, it set the industry on a course of continuous decline.[2] The disease may be the death of the Florida industry. There is something profoundly disturbing about the possibility of losing the entire orange crop (and grapefruit and tangerine crops) in a state long associated with its sunny produce. Disease-resistant varieties bring some hope, but the industry has a way to go before it reaches any safe haven. Here in the United States, a crop disease like *huanglongbing* in citrus, or wheat rust, may put a grower out of business; but in the developing world, where some families rely on homegrown grains and vegetables for sustenance, large losses can lead to hunger or even starvation. And as the world's population grows from somewhere around 7.2 billion to a projected 9 billion by midcentury, and available land becomes scarce, crop loss on large and small farms alike is increasingly unacceptable.

The first step in fighting, or even better, preventing, a problem like *huanglongbing* is identifying it. That may seem obvious, but accurate diagnosis is far from a simple process. There are hundreds of crops, each susceptible to their own set of pests and pathogens, and thousands of potential crop diseases. Consider the pathogens known to attack us humans—a single species—and then multiply. Diagnosing the ills of our food supply has been the purview of plant pathology and agricultural extension for the past century. Today, though, these fields are waning—the victims of cuts in government funding and university budgets. Both disciplines arose, over a century and a half ago, from the ashes of Ireland's Great Famine. Maybe it is time for a second round of agricultural innovation.[3]

Revelation

Early in the 1800s, late blight, or *Phytophthora infestans*, hitched a ride from the New World to the Old, kicking off one of Europe's most infamous famines. A pathogen of potatoes and related species, blight followed our favorite tubers back and forth across the continents. Sixteenth-century conquistadors, traders, and explorers first brought potatoes from their native South American soils across the ocean to Europe, Africa, and Asia. In the following centuries, tubers were cultured and tamed from their small, gnarly Peruvian ancestors into the starchy white potatoes we know today. Ireland's tenant farmers in particular became almost exclusively dependent on a single variety: the lumper. Lumpers filled bellies, paid the rent, and fed the family pig (if a family was fortunate enough to have one), which in turn provided income to survive the winter. A one-and-a-half-acre potato plot produced enough to feed a family of six for half a year.[4] The tuber helped fuel Ireland's population explosion. (Some historians suggest that potatoes may even have helped some famine-prone European states rise to become world powers.[5]) Until blight arrived. It first hit crops here in the United States and then, in 1845, it struck Europe. By that point, farmers were growing potatoes in monocrops, pumping them up with fertilizers, and had bred out undesirable but protective traits.

The potatoes were sitting ducks. So too were populations wholly dependent on the single crop. In Ireland, tenant farmers were left without food, seed for the next season (potatoes are grown from the previous year's "seed potatoes"), or hope. Artist James Mahoney, sent by the *Illustrated London News* to document the devastation, wrote that "... neither pen nor pencil ever could portray the miser[y] and horror, at this moment, to be witnessed in Skibbereen. . . . There I saw the dying, the living, and the dead, lying indiscriminately upon the same floor, without anything between them and the cold earth, save a few miserable rags upon them. . . . Not a single house out of 500 could boast of being free from death and fever."[6] At the time, the ultimate cause of the catastrophe was a mystery.

While *P. infestans* was causing mayhem across Europe, humanity was just beginning to grapple with the true nature of infectious diseases. In

1845, bad air, lack of faith, and God's wrath were all more plausible explanations for disease than microbes and fungi. That invisible life existed—oozing in pus, clinging to our teeth, or zipping around a drop of pond water—had been known for centuries. Ever since the 1700s, when Antonie van Leeuwenhoek, the draper turned lens crafter and microbiologist, became fascinated with the microscopic world (including the "animalcules" in his own semen, which, according to one text, was obtained "not by sinfully defiling himself but as a natural consequence of conjugal coitus"), humans have been aware of microbes.[7]

Yet despite their discovery, for centuries microbes remained little more than curiosities. That such tiny creatures could kill millions or cause massive crop failures was inconceivable. So when late blight blackened leaves and left potatoes rotting in farm-fields in Clare, Kerry, Mayo, and elsewhere in Ireland, there were few options other than prayer. The presence of *Phytophthora* on dead and dying plants was not in question. But was the fungus-like organism a bystander, an *ex post facto* parasite feeding off the dead plant tissues, or the cause of the disease? At the height of the outbreak, at least one astute student of fungi, the Reverend M. J. Berkeley, dared to suggest the latter. But it would be nearly two decades before German botanist Anton de Bary nailed *Phytophthora* as the causative agent.[8] It was one of the first microbes, plant or animal, to be identified as a pathogen. (Decades earlier, entomologist and microbiologist Agostino Bassi had declared that a disease killing silk worms and threatening French and Italian silk production didn't arise spontaneously but was caused by a contagious living organism, which was later determined to be a fungus.)

While recognition of the problem opened the door for prevention and eventually treatment, de Bary remained wary. According to one translation, de Bary wrote, "It will never be possible to drive the parasite *P. infestans* to extinction. . . . However, a careful selection of uninfected tubers for agriculture will be sufficient to prevent large-scale spreads of this devastating plant disease."[9]

Ironically, de Bary lived in an era when even physicians hadn't yet accepted their own role in spreading disease. Just two decades before de Bary's discovery, Hungarian physician Ignaz Semmelweis noticed high mortality rates among new mothers at a Viennese clinic and hypothesized that doctors moving from the mortuary to the maternity ward were transferring bits of dead body that infected the women. This was death not by bad air, or an angry god, but by "cadaverous particles." As a test, he insisted doctors wash their hands in chlorinated lime. Mortality rates dropped. But he could offer no explanation beyond traces of death transported from corpse to body. The physician's lifesaving discovery was a flop, coming decades before Louis Pasteur's iconic experiments disproving spontaneous generation (a prevailing explanation for why infected plants, meats, and wounds were teeming with microbes), and Robert Koch's criteria for linking pathogen to disease. Semmelweis was ridiculed. Without a compelling explanation, physicians of his era roundly rejected the idea that *they* were the source of mortal disease, and they were insulted by the suggestion that they should bother to wash their hands. For the most part, the short-lived practice of washing deadly pathogens from one's hands ceased. Nearly two decades, and many unnecessary deaths later, Semmelweis, distraught and possibly mentally ill, died in an insane asylum of sepsis (perhaps of the same ilk as the infection he tried to prevent).

Within several years, Pasteur, Koch, de Bary, and others proved what Semmelweis could not have known: that *specific* microbes caused infectious disease; that these microbes could be transferred from one body to another, or one plant to another; and that they could kill. For centuries, physicians and healers had unwittingly transmitted lethal disease, just as farmers putting away blight-infected potatoes for next season were contributing to the ensuing famine. For the first time in our history, plant pathologists and health workers could identify a once-unknowable enemy. De Bary, Berkeley, Koch, and others launched a revolution in diagnostics. Treatment and prevention followed. Koch, who identified the bacteria that caused anthrax, cholera, and tuberculosis, encouraged better hygiene, while Pasteur championed vaccination. Soon, scientists recognized fungi and other pathogens as the causes of treatable or preventable plant diseases: coffee rust, caused by a fungus; downy mildew, caused by oomycete microbes, and tobacco mosaic, caused by a virus.

This was the dawn of an age of microbiology, a century and a half ago. And here is the kicker: if Robert Koch or Anton de Bary were to walk into a plant lab or a hospital today, they would recognize the basic *diagnostic* methods being used. First, a sample is taken from an ailing patient or a dying plant. That bit of blood, drop of urine, lanced blister, snippet of leaf root is then cultured for bacteria or fungi. As the microbes are cultured, perhaps on an agar plate or in a broth, their shape, color, gases produced, and food and oxygen preferences provide a diagnostician with crucial bits of information. These clues help identify bacteria that can be difficult if not impossible to distinguish, even under the microscope. The process took days or weeks when it was first developed, and it still does. This pace may have sufficed before globalization, climate change, pesticide resistance, and a ballooning population, but it no longer does now.[10] Agriculture, like medicine, needs broad-spectrum diagnostics that can quickly identify any disease, anywhere. The need is becoming even more acute as we move away from broadly toxic treatments to increasingly specific solutions that are better for both the patient and the planet.

The Death of Pathology?

"Harry Evans is a national treasure—a veritable walking encyclopedia of knowledge on crop diseases across the world," says David Hughes, a Penn State behavioral ecologist with a vision for the future of plant diagnostics.[11] "He's not being duplicated by our society. The science-based knowledge on controlling plant-pest disease that has been happening for the last 170 years or so [since the Irish potato famine]—that's being lost." Evans is legendary. Best known for his work on diseases of the cacao tree, for the past fifty years this Indiana Jones of plant pathology has traveled the world, sleuthing and teaching about plant diseases, insect pests, weeds, and biological solutions. But today, plant pathology is becoming a lost art.

A 2012 audit in the United Kingdom suggested the field was fading quickly, as universities and colleges drop the discipline or reduce the number of courses offered, while faculty age out. The authors point out that we all understand the link between medicine and health, yet we fail to make the same linkage between plants and the food that sustains us. "Society," they say, "needs to invest in plant health in the same way it invests in human or animal health."[12] We are so focused on eating healthy food that we've neglected to consider everything necessary to keep our food healthy. In response to the current glut of PhDs in the biomedical sciences and the dearth of scientists in agricultural sciences, one biologist has editorialized, "The growing world population needs to eat and it is past due that we elevate basic, translational, and applied plant research to the priority given to biomedical research, or more boldly, to defense."[13]

Here in the United States, the number of PhD graduates in the agricultural sciences over the past fifteen years has essentially flat-lined, while the number of biomedical scientists has skyrocketed.[14] But, says Thomas Gordon, a plant pathologist at the University of California at Davis, "My perception is that we have done better than Europe and Australia, in retaining the capacity to train plant pathologists."[15] Still, that capacity is shrinking as programs like those at Cornell University and North Carolina State University, once giants in the field, are merged into other departments, and as plant pathologists retire. "Most voters know and care very little about agriculture," says Gordon, who believes the loss is, in part, the result of dwindling public financial support. "Agriculture is largely seen as just another large industry, and so should have no special call on public funding to address their problems." We clamor to break away from industrial, overly processed foods and we increasingly demand organics, yet we fail to understand the difficulties faced by farmers, particularly smaller growers. Imagine losing infectious diseases experts while facing a newly emerging disease like Zika. We will suffer the loss of those who help keep agricultural crops productive and disease-free. It is time we took notice.

And we aren't only losing practitioners—disease experts who can finger the culprit when leaves start to turn brown or fruit is yellowing. Diagnostic materials, including images and descriptions of plants and plant diseases, are being increasingly locked away behind paywalls. In Europe, Africa, and elsewhere, extension services are becoming too costly or are privatized as governments struggle to divvy up shrinking funds. Essential diagnostics services are unavailable to those most in need: the small-scale growers, the less affluent farmers, and those in remote regions and developing nations.[16] "What is lost," says Hughes, "is a process of studying the problem. Everyone wants a silver bullet." We want quick fixes, when sometimes a deeper knowledge of a particular pest or how spores are transported on the wind may provide longer-lasting solutions. The trend is disturbing enough to Hughes that he is willing to risk his career in the attempt to develop a remedy.

Hughes and colleague Marcel Salathé, a digital epidemiologist and former app developer, had an idea. They would build a *free* digital library of plant-disease images and information, and fill it with voluntary virtual librarians experienced in plant diseases from around the globe. That would be step one. Step two would be to automate the whole process. "At the same time we're losing investment in people like [Harry Evans], we have a lot of people running around with computers in their pocket," says Hughes. Millions of people, from small growers in Africa to backyard gardeners in the United States and wheat growers in Europe, carry mobile phones. "So there's a possibility of a second-best solution. Connect these individuals so we have a social network. There is a *lot* of collective knowledge." If that knowledge can be corralled, digitized, and eventually shared, Hughes imagines a day when a grower stymied by a newly infected crop might point, shoot, and have a diagnosis within minutes or even seconds. Any disease diagnosed anywhere, anytime, for free. Salathé sees an even broader application of the technology, writing, "if you can do this with plant diseases, you can do this with human diseases as well."[17]

Hughes's "day job" at Penn State is hunting down fungus-infected "zombie ants" in far-flung regions of the globe. But his vision is to reimagine a basic service traditionally provided by governments and universities: agricultural extension. He acknowledges that this work "doesn't get you anywhere [professionally]. For me and Marcel to build this went against a lot of perceived wisdom and advice. We were told not to do this—that it was career suicide . . . but anyone with half a brain cell knows that if we are going to feed 9 billion, maybe making all the knowledge of how to make more food available is a good thing. Not doing it was the morally corrupt thing, so Marcel and I happily took the road less taken because it would lead to a better world for all of us."

Extending Knowledge

For much of my early career, I traveled from one so-called land-grant university to another, crisscrossing the country in my pink Corolla (packed with books, cats, and eventually dog, husband, and kid). I went from New York to California, back to New York, then to Rhode Island, then North Carolina, before setting roots in the rich soil of the Connecticut River Valley, near the University of Massachusetts. I never paid much attention to agricultural sciences, yet each of the universities where I worked originated as a land-grant institution.

First envisioned by Illinois College professor Jonathan Baldwin Turner, and sponsored by Vermont congressman Justin Morrill, these universities were designed to offer public education for all, no matter their social class. Signed into existence in 1862 by President Lincoln—in the midst of the Civil War—the land-grants taught agriculture, military tactics, mechanical engineering, and classical arts to those who couldn't afford private colleges. These institutions were the place for those raised on the farm and intending to return to the farm, if not physically then intellectually. Much of our nation's approach to agriculture emerged from their halls. The land-grants developed insights into plant nutrition; new varieties of corn, wheat, and apples; pest controls and pesticides; disease controls and resistant varieties. They were the home of the Green Revolution, with all its ups and downs. Land-grants continue to provide critical support for one of our most important endeavors: growing food. One invaluable service is extension. Starting up an apple orchard? Curious about scrapie in sheep? Have a new pest issue? Tracking blight? You can e-mail your local extension agent; or send a soil sample for testing or a leaf for diagnosis; or attend an Integrated Pest Management field walk. Extension translates the latest science for growers, from the most effective pesticides to agroecology and permaculture—approaches designed to reduce chemical input.

Better knowledge for better food: a concept put into practice in the midst of the Irish potato famine. As potatoes rotted in the 1840s, George Villiers, the Earl of Clarendon and Lord Lieutenant of Ireland, sought to prevent "recurrence of the great calamities" by sending a handful of agricultural lecturers into the countryside. Their task was to "diffuse that agricultural knowledge upon which the very safety of the country now depends." Lord Clarendon believed that teaching farmers *how* to grow more-robust crops and encouraging them to grow food other than the potatoes could prevent another disaster.[18] At the very least, if potatoes failed, there would be another food crop. The first ten or so agents sent into the stricken countryside were successful enough. Soon this kind of knowledge sharing spread around the globe. Just after the turn of the twentieth century, it took shape here in the United States as a cooperative venture between the US Department of Agriculture and land-grant universities. The programs would both produce "practical applications of research knowledge" and share this knowledge with the public, from the Midwestern wheat growers to the orchards in the Northeast to citrus growers in the South to the backyard gardener.

But over the past century, knowledge has become an agricultural "input," akin to large farm equipment, fertilizer, or water. Agribusiness has the resources to pay for this knowledge, whereas new science has become less accessible to small growers, particularly those in far-flung regions. "Since commercial farmers can derive direct financial benefits from these inputs, there is a trend towards the privatization of the extension organizations," observed the Food and Agriculture Organization of the United Nations back in the 1990s, ". . . with farmers being required to pay for services which they had previously received free of charge. This trend is strong in the North, and there are examples of it beginning in the South."[19] Here in the United States, small growers—depending on their region—are still able to call upon extensions for free, unbiased scientific advice, even as funding has become increasingly scarce.[20]

"We can have a new type of extension," says Hughes. "We shouldn't be thinking we are stuck with this model we've used for the past thirty or forty years. We can change. Extension began in my hometown of Dublin in the Phoenix Park where I walked as a kid and was implemented in the fields of Connemara where I once lived. I see no reason I can't accelerate its change 170 years later to fit a twenty-first-century world." Hughes wants to put a Harry Evans in the pockets of growers, from those tending "community gardens in Brooklyn to smallholder farms in Burkina Faso."[21] The goal is diagnostics that zip from satellite to cell tower to the palms of growers just about anywhere.

Smartphone Village

Nearly 100 years after the emergence of extensions here in the United States, and 170 years after the idea's beginnings in Ireland, Hughes and Salathé envisioned PlantVillage.org. The site offers images, information, and experts. Send in a photo, ask a question, get an answer. The ultimate goal is to turn the lion's share of diagnosis over to the vast computing power of the Internet, while saving the trickier cases for the dwindling supply of experts like Harry Evans. The site went live in 2013. By 2016, PlantVillage had acquired over 100,000 images covering 155 crops and 1,800 diseases. More than two and half million visitors, roughly a third of them from developing countries, have submitted photos, asked or answered questions, or read the content for solutions. How do you deal with grubs on basil plants, ants invading banana plants, fungus on watermelons? Instead of the ten agents "possessing sound practical knowledge" of Lord Clarendon's day, there are hundreds or thousands or perhaps someday, millions, sharing knowledge—without leaving their laboratories or farm-fields. Currently a grower can snap a photo, post it, and within minutes a diagnosis and solution might be posted by contributors. As for images, Hughes and Salathé are aiming for millions over the next few years—"a bit of a moonshot," he admits. If growers in remote regions of Africa or India or Brazil don't yet own smartphones, chances are they will in the near future. There are currently more mobile phone subscriptions than there are people on the planet. A couple of years ago, roughly 75 percent of those living in sub-Saharan Africa owned mobile phones. In Nigeria and South Africa, almost everyone owns a phone.[22] Although a little less than half of all subscriptions are smartphones, by 2019 the number of smartphone users is predicted to rise to over 6 billion.[23] Growers will have smartphones; and Salathé and Hughes are counting on social networking and crowdsourcing for better diagnostics.

At first glance, PlantVillage looks intriguing, but not like a site capable of saving food crops around the globe. "I would not expect miracles," says one extension agent when I show him the site. Another, Margaret Lloyd, the small-farms advisor for the University of California (and strawberry expert, whom we met in chapter 2) who has sat in on discussions about this kind of approach, points out the difficulties with photo quality (which Hughes says they've solved with an algorithm), regional conditions that might come into play, and a lack of commitment by the experts.[24] "Diagnosis can be tough even for trained plant pathologists because it can be an investigative process that takes the right questions and good training to home in on the probable cause," she adds. "There are very simple ones that can be diagnosed by a simple photo, but most farmers/growers learn these within the first few years of growing." Even so, she's hopeful about the online endeavor: "I would be thrilled to hear of this working; it's a noble concept."

Local farmers would like to see the site in action as well. Ryan Voiland of Red Fire Farm here in the Connecticut River Valley, one of the larger CSAs, with over 1,300 subscribers, says he'll send a sample to extension for diagnostics, but, with testing costs running $50 a sample, he does so only sparingly. A system providing easy, accurate visual diagnostics would be welcome. At the very least, it would provide a more informed guess. But as Lloyd points out, it is difficult to get scientists, who are under pressure to teach, win grants, do research, and turn out publications, to contribute freely to sites like PlantVillage. Academic bean counters don't count contributions to these kinds of traditionally non-academic websites even if they are not-for-profit, so there is little incentive. At least not currently. It's a problem that Hughes acknowledges, but he's optimistic about the next generation of academics: "I think the younger generation is much more altruistic and less focused on their own CVs and their own rewards. This is the future, moving towards more integrated spaces." And Hughes and Salathé aren't counting solely on human brain power. They have bigger plans. They want to *teach computers* to diagnose plant diseases.

A Brain in Every Phone

World War II was a massively destructive human endeavor. But without the energy and knowledge channeled into producing more-efficient methods of killing, it's possible that we might not be carrying wallet-sized computers in our pockets today. The war years brought us industrial-scale production; the packaging of penicillin; radar; pressurized airplanes; and a whole slew of other mid-twentieth-century innovations that surround us today. The need for more-accurate ballistics inspired John Mauchly and John Eckert Jr. at the University of Pennsylvania to create the world's first digital, *all-purpose* computer. Booting up for the first time in 1945, ENIAC (Electronic Numerical Integrator and Computer) never did, in the end, direct missiles in World War II, but its capacity for massive high-speed calculations and its ability to be programmed were revolutionary. ENIAC was a behemoth weighing nearly thirty tons and spreading out over 1,800 square feet; its descendents now accompany us to the market, into the classroom, out to dinner, and into the field. ENIAC and its immediate successors like EDVAC and MANIAC did what was asked of them, solving problems posed in a particular way. But could they do more? Could machines made of wire and vacuum tubes, with electricity coursing through their metallic veins, learn to think like us, or maybe even better? British mathematician Alan Turing thought so, foreseeing what is now known as artificial intelligence (AI). Turing helped break the German Enigma encryption during the war and was influential in the early development of computers. Like many of Turing's contributions, AI was an audacious proposition.

Turing wondered if the new computing machines could learn as a child does, and *if* they could learn, what would that really mean? Could a machine, an artificial brain, essentially filled with "the best sense organs that money can buy," be taught to understand?[25] He posed the question in 1950. Unfortunately, the brilliant mathematician who pondered unrestricted machine intelligence lived in a time of tragically limited human understanding. Arrested for homosexuality in 1952, and offered a Hobson's choice of avoiding prison only by consenting to "treatment," Alan Turning underwent the unconscionable procedure *de jour* known as "chemical castration." Two years later, at age forty-one, Turing swallowed cyanide. In the fifty years since Turing's death, inspired by his visionary essay "Computing Machinery and Intelligence," computer scientists have programmed computers that learn how to play checkers, recognize patterns, and make sense of "big data." Computers have beaten humans at chess and more recently at go, a vastly complex game of strategy and intuition that was one of the last holdouts in the human domain.

Now computer scientists, collaborating with the Smithsonian Institution, have built a smartphone app called Leafsnap that can recognize leaves from the northeastern United States and Canada with a snap of a photo.[26] It is just one application of pattern recognition. Offering Leafsnap for free and harnessing the power of the crowd, by 2016 the site could identify all 185 northeastern US tree species.[27] I recently downloaded a similar app for identifying flowers. Holding this collective brainpower in your hand is a thrill. And this kind of pattern recognition is just one branch of machine learning. But Hughes and Salathé are delving deeper into the electronic brain. "Leafsnap just finds the edges," explains Hughes, comparing the app to their own PlantVillage app, "and uses that to determine the plant species. We look at the leaf surface and use color, shape, and orientation to determine disease. It is the difference between seeing an outline of a person and knowing it is a man, compared to seeing the features and recognizing it is Uncle Bob (friendly) versus Hannibal Lecter (not so much)."

To make PlantVillage work, Hughes and Salathé have turned to deep learning: developing computer algorithms that can deal with high-level abstractions. One product from Google is TensorFlow, an open-source machine-learning software library. In 2012, the project announced it had connected 16,000 computers together, creating a brain-like network. When loosed on some 10 million randomly chosen digital images, the network *taught* itself to recognize thousands of objects. Thanks to our penchant for posting photos of cats, one of the better-known objects to emerge in the learned digital brain was the image of a cat. Whenever we look something up on Google or use Gmail, TensorFlow is there. Hughes, Salathé, and Sharada Mohanty (a graduate student in Salathé's lab at the time) used a similar kind of deep-learning framework called Caffe on a set of 53,000 images of fourteen crops with a combination of twenty-six different disease categories. The computer was able to assign the 26 diseases with 99 percent accuracy. "That is really good," says Hughes. "There is *no* way a person—a plant pathologist or any person on the planet—could do that. While you may be an expert on tomatoes, that doesn't necessarily mean you know about fungal disease on grapes. And if you know about viruses, it doesn't mean you are good at classifying grape diseases caused by fungi." Since then, the pair have released their image database on a competition platform called crowdai.org where computer scientists around the world competed to get the best disease identification results, using their own approaches. The goal, says Salathé, is to find out which approach works best. Any results must be open-source and available to the public.[28] "It was very satisfying that other teams got a greater than 99 percent accuracy, which is extraordinary," glows Hughes. "It shows that a village is always better."

Of course, a diagnosis is only as good as the data. Images need to be clear. And the diagnosis has to be confirmed—a challenge that Hughes and Salathé are well aware of. "Once we go to a completely different data set (a few images we took from the Web, from sites with a high reputation), the accuracy reduces substantially," says Salathé. One remedy is to diversify the entire data set—which the pair have recently begun to do. "The vast majority of the current disease-training set, 150,000 images and growing," says Hughes, "have all been identified by an expert who either infected the plant or knew what it was. But we can't expect everything to be so well studied, especially in places like sub-Saharan Africa. We are going to have to ground-truth it"— that is, test results in the field or in the lab. Recently, the team worked in Tanzania with the International Institute of Tropical Agriculture and one of its lead scientists, James Legg, to do just that. (IITA is part of a global consortium of research centers devoted to agriculture and sustainability.) Hughes is now working with hundreds of scientists around the world to get images and double-check the system's diagnoses. Given the scope of the project, it's a tall order, particularly if the option for definitive diagnosis is the age-old practice of laboratory culturing. But even in the laboratory, computers are influencing science in ways their innovators could never have imagined. We are entering an age where *fourth*-generation DNA sequencing is poised to revolutionize disease diagnostics. Just as computers morphed from room-sized to desktop-sized to wristwatch-sized, DNA sequencing has increasingly sped up, downsized, and become more affordable. How might one diagnose a cast of hundreds or thousands of known diseases—cheaply, quickly, and accurately? Or, for that matter, flag an emergent unknown disease? Fourth-generation DNA sequencing can do this.

"We are pretty far away from lab-on-a-chip diagnosis of *any* kind of organism," says Hughes, referring to efforts to move from a roomful of laboratory equipment to a computer-chip-sized operation. Instead, his lab uses a device produced by Oxford Nanopore, a self-described "disruptive" tech company (more on this in the next chapter), that connects by USB to a laptop and spits out DNA sequences in minutes—which means that a bacterium or virus can be identified in minutes. If these new sequencers are successful, rather than sending out a snip of stem or a vial of blood, diagnosticians could soon hold the answer in their hand. The parent company is already developing a mobile phone version. The devices are being put to the test in the field, on the orchard, and in the hospital, and results are just now appearing in the scientific literature. Diagnoses range from Ebola to Zika to the more mundane but intractable antibiotic-resistant urinary infection. This move toward rapid DNA diagnosis, the topic of the final chapter, may chart a whole new course in disease diagnostics. It could be a key for innovators like Hughes, Salathé, and others, devoted to saving our food, and for ER physicians, rural doctors, and health care workers in the field, devoted to saving our lives.

CHAPTER 8

Know Thine Enemy: The Future of Diagnostics

When my husband spiked a fever a few weeks after a tick bite (the tick was removed early, no rash), he went to the doctor just to be safe. The physician noted symptoms that pointed to a flu or respiratory virus. Nevertheless, Ben came home with a prescription for a single high dose of doxycycline—for Lyme disease. He had given blood for a Lyme test, but the drugs were prescribed during the visit, before any lab results could be available. The blood tests detect antibodies made against the Lyme bacterium and can be turned around in days. But the tests are most accurate only weeks after an infection takes hold and, depending on the timing of the test in relation to exposure, a response may indicate current or past infection. Culturing the bacterium is difficult, which leaves physicians in a precarious situation—to treat prophylactically in a region where Lyme is endemic or to wait. To let Lyme go is to risk a more intractable infection.[1] Without a sensitive office test, the proactive option often seems like the best one. My husband took the antibiotic; a few days later he was informed there had been no active infection.

Throughout this book, we've seen the consequences of antibiotic misuse or overuse in the hospital, at the doctor's office, and on the farm. After a century of chemical warfare, the bugs are striking back. In 2016, for the first time, researchers identified an *E. coli* bacterium resistant to colistin, a "last resort" antibiotic that fell out of favor nearly fifty years ago because of its toxicity, but which has since been reenlisted in response to resistance. The bug itself can be cured with other antibiotics, but the mere *presence* of a resistance gene on a bit of DNA that is easily shared with other bacteria is scary. The "pan-resistant" bug was cultured from a patient with a urinary tract infection here in the United States, where resistance claims at least 23,000 lives annually—and more when those who die from complications caused by resistant infections are included. The incident set off a flurry of doomsday articles and interviews.[2] If new solutions to resistance and infection are not found soon, the number of deaths both here and globally (where hundreds of thousands die from resistance each year) is projected to rise into the millions by 2050.[3] Resistance not only kills but it is also expensive, racking up roughly $20 billion a year in health care costs in the United States alone. It affects families, work, and available hospital space.[4] Resistance looms large, as does, for hundreds of thousands of recipients, the collateral damage wrought upon their microbiome by broad-spectrum antibiotics, which create a wasteland ripe for invasion by opportunistic bugs like *C. diff.*

Accuracy—targeting the harmful, while sparing the beneficial—is one way out of our current conundrum. We should aim, writes physician and microbiome pioneer Martin Blaser, "to develop truly narrow-spectrum agents, each ideally targeting a single pathogen."[5] Phages can do this. So can bacteriocins and other narrow-spectrum agents. But these treatments require improved diagnostics. While microbiologists, pathologists, and diagnosticians have become very good at identifying whatever ails us, the process for a definitive diagnosis takes time—often days or weeks. In our rapid-paced, Internet, next-day-delivery, high-tech world, we diagnose disease with what is essentially nineteenth-century methodology. Culturing—growing out bacteria in test tubes, petri dishes, and flasks—remains the gold standard. If we are to save our antimicrobials and our lives, we need twenty-first-century diagnostics. We need tools that can be used at the bedside, in the office, and in the remote rural hospital. And we need diagnostic tools that can read into a bacterium's genome and tell us whether it remains sensitive to antimicrobials. And all this must be affordable to all health care providers, because otherwise, those providers have to make a gamble. Sometimes that means dispensing broad-spectrum antibiotics as a kind of catch-all cure.

Although doctors are becoming more conservative with their prescription pads, probably all of us have a story about receiving antibiotics before a final diagnosis. It's likely that your little ones won't be given antibiotics for that ear infection as easily or as often as mine were. Yet even in 2016, when my daughter was home from college and coughing her way through spring break, I sent her off to the doctor's office and she returned with antibiotics. Surprised, I asked if she'd been tested for anything. She hadn't. There'd been a look down the throat and up the nose. About one-third of all outpatient antibiotic prescriptions were recently deemed "inappropriate."[6] Roughly 50 percent of patients seeking respite from a cough, a symptom of acute upper-respiratory-tract infection, still walk away with antibiotics.[7] Even a quick test to distinguish virus from bacterium (now available in Canada) would have helped.

The problem isn't a *lack* of diagnostics per se. According to a recent market study, identifying infectious microbes is a $16-billion industry and growing.[8] One portion of those billions goes to "traditional identification" and testing for antimicrobial sensitivity—or culturing. Another chunk of money is spent on tests that capitalize on antibody-antigen reactions or probe for specific DNA sequences. Many of these tests have been around for decades. More than twenty years ago, a rapid diagnostic antibody-antigen test informed me that I was pregnant with our son Sam. You can buy a rapid HIV or gonorrhea test kit at Target. A repeat offender for strep, my daughter is a rapid-test veteran. The drill is a quick throat swab, a yea-or-nay diagnosis, with an antibiotic chaser (or not), followed by a culture for the ultimate confirmation. Now there is even a rapid test for Ebola. But there are over 1,000 *known* human pathogens, and most commercial rapid tests focus on some of the more common or troublesome infections worldwide: multi-drug-resistant staph, TB, *C. diff.* Less common infections are left out of the rapid-diagnostic loop.

And as good as existing rapid tests are, some are too sensitive while others not sensitive enough; or they are too expensive for smaller hospitals and offices; or they cannot diagnose the precise strain of the infection. A test may not distinguish between a harmless bacterium and, say, a toxin-producing cousin, or one that resists penicillin. Even existing molecular, DNA-based tests are no panacea for detecting resistance. "The molecular targets for certain antimicrobial resistance determinants are available," explains Joseph Schwartzman, medical director of microbiology at Dartmouth-Hitchcock Hospital, "but are not yet comprehensive, nor are all resistance phenomena explained by single genetic elements."[9] In other words, a molecular test that probes for a single gene or even a few isn't sufficient. Many of these diagnostics still require a day or two to produce results—sometimes, a day or two too long. Most doctors encourage follow-up culturing. What is lacking is a *robust* diagnostic system. We need tests that can tell a bedside physician not only whether a patient has an infection, but how a bug will respond to antibiotics. Or whether the strain detected in a patient is friend or foe.

As we enter an era of the gene and genomics, rapid molecular diagnostics are moving, well, rapidly. There are tests approved by the FDA, laboratory tests that aren't yet FDA-approved still in need of extensive validation, and tests that remain designated as research-use-only. Diagnostic laboratories of various sizes find that they have options according to the patient populations they serve and the number of specimens received for various diagnostic tests, says Schwartzman. In other words, it's complicated. And what may be available at one hospital or for one patient population may not be available for another, particularly in smaller, far-flung regions.

Bringing effective diagnostics to remote areas is increasingly important as we humans become ever more mobile. Consider the world we live in, where a pathogen can chase the sun, traveling from east to west, or north to south, in a day. In 2014, the United States welcomed some 75 million international visitors to her shores (about half from Canada and Mexico, but the rest from farther afield and overseas), almost double the number from a decade earlier.[10] In 2015, over one billion tourists made their way from one destination to another worldwide.[11] People provide the transport for unseen hordes of hitchhiking microbes: Ebola, influenza, and now Zika (primarily mosquito borne, Zika can also be transmitted sexually). Zika and Ebola are harbingers. Diseases once considered exotic will continue arriving on our shores and in our airports. More will emerge, eventually showing up in hospitals and doctor's offices by patients seeking a cure.

Instant Gratification

"We certainly do wish we could do more rapid diagnosis of pathogens," says physician Kathleen McGraw. "There are too many times when we have to use broad-spectrum antibiotics, just to cover many possibilities until we get results."[12] McGraw is the chief medical officer of a rural community hospital in Brattleboro, Vermont, that services a population of about 55,000 spread across a couple dozen towns and villages. "I'm thinking about sepsis as a big issue for this. What if there were rapid tests for some of the encephalopathies and other kinds of serious and life-threatening infections, where we are treating people with toxic antibiotics or antivirals?" Or what if there were affordable, dependable, "point-of-care" bedside tests for MRSA? When a patient comes in and might have MRSA, blood is sent to the lab, where definitive diagnosis takes an average of two or three days, says Christopher Appleton, medical director of pathology at McGraw's hospital.[13] By then it's possible that the bug might have spread, contaminating surfaces or health care workers or a patient's roommate. Conversely, says McGraw, "If a patient has *ever* tested positive—anytime they reenter the hospital, they are treated as if contaminated, until proven otherwise."

"We still culture a whole bunch of stuff—the full boat," says Appleton. "We culture it when we don't have rapid tests for it." Appleton points out that while they use rapid tests for infections like *C. diff*, strep, and a few others, they still culture afterwards, no matter what the test says. Moreover, they routinely culture fluids from joint-replacement surgeries: "We culture it for *Proprionibacterium*, the same that's on people's skin, responsible for acne." The bacterium, which normally coexists with us peacefully on our skin and in our mouths and ears, is an emerging pathogen in orthopedic and implant surgery and can cause serious problems. This particular bug, which can thrive with little or no oxygen, is slow; which means that growing out a culture can stretch out over several days or even weeks. "We are still stuck in the old days. Microbiology in a community hospital still takes a while," says Appleton. Brattleboro isn't unique. There are over 1,800 rural community hospitals in the United States.[14] Lacking available rapid tests, smaller labs like Appleton's send samples for some of the less common tests to larger academic hospitals or clinics. More common tests are done in-house—it's a balance between time and cost. "We have a good system where we send a lot of material to the University of Vermont in Burlington, and if they don't do the test they send to Mayo. It's quick—we have couriers who come every day." With a turnaround time of twenty-four hours or less, it beats culturing. Quick, but not bedside, while-you-wait, pregnancy-test rapid.

Having a baby isn't exactly the same as having an infection, but the rapid test for pregnancy paved the way for "dipstick" disease diagnostics. The pregnancy test is the iconic rapid test. Within minutes of testing, you know that either you are or you aren't. Once a fertilized egg implants into the uterine wall, cells begin releasing human chorionic gonadotropin or hCG (one of the better-known pregnancy hormones). Some of that hCG is excreted in urine. Concentrations rise as early pregnancy progresses, peaking several weeks after implantation. Many dipstick tests on the market are most accurate a day or so after the first missed period, or roughly two weeks after implantation—give or take. The pregnancy test wand contains antibodies specific for hCG antigen (along with less specific antibodies able to bind to *those* antibodies). When the antigens and antibodies combine, the test announces the impending pregnancy by a blue line, or pink, or a plus sign; some even spell out "Pregnant." Developed in the early 1970s as one of the first dip-and-read color diagnostics, the first rapid pregnancy test was a failure. But the concept prevailed.[15] A few years later, a so-called ELISA test (enzyme-linked immunosorbent assay) was developed to detect malaria in patients in the field in Tanzania, demonstrating the power of combining antibodies, antigen, and color. An elegant and sensitive assay, the test opened diagnostic doors. Rapid detection of hormones, toxins, antibodies, and a host of infectious diseases, from HIV to hepatitis to *C. diff* and *C. diff* toxins, would follow. ELISA tests revolutionized medical diagnostics, releasing diagnosis from laboratory-bound equipment and compressing diagnostic times from days or hours to minutes. "Looking back," writes one of the developers, "I continue to be amazed about what came out of our (too) early attempt to develop a simple pregnancy color test!"[16]

McGraw's hospital makes use of over 100 rapid tests—each with a different diagnostic target. Many are ELISAs, most are for health-related indicators other than infectious diseases. There are tests for drugs, sperm, pregnancy, kidney and liver indicators, and glucose. A dozen or so are aimed at pathogens, including tests for specific respiratory viruses, strep, or *C. diff* antigen. With infections such as C. *diff* and MRSA on the rise, particularly in hospital-associated infections, or HAIs, screening can not only prevent heartache but also reduce hospital costs. Hospitals also compile antibiograms, or annual tallies of antibiotic resistance in common pathogens, that help doctors make more-informed decisions. Collectively across hospitals, HAIs rack up tens of billions of dollars in treatment costs annually. Depending on the infection, a resistant bug may double treatment costs. Appleton says that for an infection like *C. diff*, the ELISA is a first step. Though the test is sensitive enough to indicate the presence of the bug, it can't differentiate between a patient who is simply carrying *C. diff* and one who is infected with a toxin-producing strain: "If it's positive, we can go to molecular testing for the specificity part of it. We are just getting up and running now." A second swipe at *C. diff*, using a DNA-based test, provides more-detailed information. ELISAs enabled health care workers to dip and diagnose, but as Appleton points out, the tests may require follow-up. Sometimes the antibiotic comes before an accurate diagnosis, as with my husband Ben.

As fans of crime shows know, a perpetrator's DNA on a glass or hairbrush or lipstick can unravel the most meticulously planned crime. While early DNA analysis required gobs of sample, now the smallest trace of DNA can be amplified by millions or billions for diagnosis. A few molecules or even a single gene can be picked out by a DNA probe and copied—over and over again. It's like the old shampoo commercial where two friends, telling two friends, telling four friends, and then eight friends, exploded into to dozens before our eyes. This kind of PCR (polymerase chain reaction) technique amplifies DNA in a similarly exponential way. A lab like Appleton's might send samples off to the University of Vermont or the Mayo Clinic, which have the technicians, equipment, and dedicated laboratory for molecular analysis. If they are lucky, PCR might confirm infection right down to its antibiotic resistances within hours. The added information can be a lifesaver. But there's a catch. PCR amplifies *specific* regions of DNA: a gene sequence specific for toxin producing *C. diff*, or chlamydia, or a gene that enables a bug to resist an antibiotic. It is a powerful test, but you need to know what you're looking for.

And there are other caveats. As with almost any diagnostic, test errors can lead a diagnostician to conclude infection when there is none, or vice versa. These false positives and false negatives can be a drawback of some PCR tests. In particular, one of PCR's strengths—sensitivity—is also one of its greatest weaknesses. The test is so sensitive that DNA that may be wafting in the air (or on a lab bench or on a clinician's gloved hands or in a pipette) might contaminate a sample. Should a segment match up with the PCR target, it will be amplified, and a test may indicate infection when there is none. In 2006, when physicians and health care workers at Dartmouth-Hitchcock found themselves coughing for weeks on end, the infectious disease specialist wondered if whooping cough might be sweeping through the hospital. Concerned about patients and workers, and tipped off by clinical signs suggesting whooping cough, the hospital used a rapid PCR test for confirmation. Back then, the test was a relatively new diagnostic tool. The results sent the hospital down a whooping cough rabbit hole: over a hundred cases were diagnosed with the infection; over a thousand health care workers were treated with antibiotics; and over 4,000 workers were immunized within days. Follow-up samples were sent for culturing, but the suspected culprit, *Bortadella pertussis*, is a slow grower. After over a week of culturing, no whooping cough bacteria were found. Follow-up blood samples turned up negative for antibodies. Additional PCR tests by the CDC also turned up negative.

There had been no outbreak. But Dartmouth-Hitchcock wasn't alone that year in its initial misdiagnosis. Later in 2006, a Massachusetts hospital also incorrectly diagnosed whooping cough infection based on PCR, concluding that dozens of health care workers were infected. Two years before that, in Tennessee, an infant's diagnosis using PCR set off testing and treatment of symptomatic community members who'd been in contact with health care workers. In the end, culturing revealed that only the infant had been infected.[17] In a *New York Times* article about the Dartmouth-Hitchcock "epidemic that wasn't," one infectious disease specialist commented, "It's almost like you're trying to pick the least of two evils." That is, you are faced with an infection that can sweep through a community like wildfire. So do you wait for a culture that can take weeks, or use a molecular test that can be wrong? The test isn't a panacea, but with a disease like whooping cough, rapid diagnosis is critical. That was over ten years ago—ages in our high-tech, molecular world. PCR is becoming more foolproof—increasingly accurate and more difficult to contaminate. Still, despite improvements, the CDC recommends confirmation by culturing.[18]

Tuberculosis is another notoriously slow-growing bug, and culturing can take weeks—time enough for a patient to infect plenty of others. The GeneXpert is designed to prevent that. Developed a decade ago by medical product maker Cepheid, the small, box-shaped machine uses PCR to detect resistant TB infections within hours.[19] It uses cartridges so that different steps in the process, normally done by a technician, are completed within an enclosed space, reducing opportunity for contamination or human error. It's more effective—and more costly—than culturing. But the test offers an unprecedented opportunity to pursue TB infection, which has become entrenched in regions of Africa and India and elsewhere. So, with substantial supplemental funding from global organizations including the Gates Foundation, 200 machines and millions of TB test cartridges were sent to Africa, Eastern Europe, and Asia. According to the World Health Organization, by reducing time for TB diagnosis the GeneXpert has helped save over 60,000 lives.[20] The machine, says medical microbiologist Justin O'Grady, "is exemplary for rapid diagnosis," bringing PCR to far-flung regions where sending samples by courier to a University of Vermont or Mayo Clinic just isn't an option.[21]

Pathologists like Christopher Appleton are waiting for the day when these kinds of tools are affordable for a small hospital with a low overall infection rate. A cartridge, test-card, or plate may not be all that costly, but the machines can run tens of thousands or more, bumping up the overall cost of a test. (The company's investor relations analyst, in an email writes that GeneXpert costs more per unit, or per test, than devices from other manufacturers, but says the expense is balanced by savings from fewer infections in hospitals.) Meanwhile, the FDA continues to approve dozens of new tests each year.[22] So hospitals with limited resources must pick and choose if they want more-advanced testing. For in-house testing other than ELISAs, Appleton relies on a Vitek machine. Developed by NASA to ensure that astronauts returned sans extraterrestrials, the device has since became available to hospital labs. Sold by the French company bioMerieux, the machine uses neither ELISAn or PCR but instead distinguishes one microbe from another biochemically, testing cultures against a gauntlet of metabolic and chemical challenges, including antibiotics.[23] The company claims to identify up to 100 organisms. But prior to hitting the Vitek, samples must be cultured for anywhere from eight to twenty-four hours. *Then*, says Appleton, "Vitek can identify it pretty darn quickly," conceding wistfully that microbiology always takes a while. Even with Vitek, ELISAs, and PCR diagnostics becoming increasingly available in our hospitals and doctors' offices, each is not enough. Some are speedy but not accurate. Others are accurate but too expensive. All have a limited breadth of diagnostics. There is still no magic dipstick, wand, or diagnostic capable of identifying *any* pathogen, in any tissue, in short order.

How do you diagnose a patient with symptoms that just don't fit the "usual suspects"? When Zika exploded into our lives, health officials raced to monitor its advances, but rapid diagnostics were unavailable. Now there *are* diagnostics, but they are in limited supply. In the midst of the 2016 outbreak, samples sent to the CDC for testing could take more than three weeks, and perhaps longer for a diagnosis, according to their website. Then there are all those "healthy carriers." Rapid diagnosis isn't just for the obviously ill, but also for the seemingly healthy, particularly in cases where people who feel fine can spread infection. We need rapid, accurate, sensitive, and broadly diagnostic tests—now. At least one option, MinION, has been making the rounds of research laboratories around the globe, and it is looking promising.

Diagnosing in Code

The dearth of diagnostics, combined with mounting global anxiety over antibiotic resistance, is driving private investors and governments alike to take notice. And recently, the money has begun flowing: in 2012 Qualcomm, a developer of mobile technologies, announced the Tricorder XPRIZE, promising $10 million to the first team to produce, by 2017, a *Star Trek*–inspired device that brings health care "to the palm of your hand"; in 2015, the National Institutes of Health distributed $11 million in first-year funding for diagnostic tools to identify resistance in selected bacteria within three hours or less; and that same year, the EU Horizon Prize promised a million euros for a cheap, fast, easy-to-use and minimally invasive test to determine whether a patient even needs antibiotics. One new technology is already getting plenty of field-testing, some of it in medical diagnostics: nanopore DNA sequencing. DNA tests like PCR use specific DNA primers or probes. Nanopore sequencing requires neither. Just as next-generation sequencing technologies are revealing the microbial world within and around our bodies, they may also lead us to the Holy Grail of disease diagnosis: a way to identify any known pathogen, anywhere, within minutes.

Justin O'Grady devoted years of research to PCR-based diagnostics.

While PCR has had plenty of success, he says this approach hasn't really taken off because "PCR can only really give you the answer for the question you've asked. You can only find what you're looking for." Now he, along with other early adopters or alpha-testers, has been working with a device called MinION, developed by Oxford Nanopore. MinION is a DNA sequencer, and researchers like O'Grady are putting the device through its diagnostic paces.[24] O'Grady says the appeal is its speed, its potentially lower cost, and its expansive capacity to identify microbes—in fact, *any* known sequenced microbe. There is no need to amplify or target a specific preselected DNA sequence. MinION reads *each* DNA base as a single strand of the nucleic acid passes through a vanishingly small hole, or nanopore. It is like reading out any random word in a book rather than reading preselected words that must be transcribed letter by letter onto another page.

Nicholas Bergman, head of genomics at the Department of Homeland Security's National Biodefense Analysis and Countermeasures Center, explains the key difference between MinION and diagnostic machines like GeneXpert: "They are all nucleic-acid-based. There are really two categories of method—agent-specific and agent-agnostic."[25] Nanopore sequencing is agent-agnostic. PCR-based tests are good at asking, "Is pathogen X in my sample?" but they are ill-suited to the broader question of "What's in my sample?"

Says McGraw, "This concept of 'what is in my sample' is the dream of physicians when faced with a patient with 'FUO,' or fever of unknown origin." When an undiagnosed patient's condition continues to deteriorate, "infectious disease consultants are called, and patients interrogated for whatever they could possibly have been exposed to, but we can only test for things we guess might be a possibility."

In the case of my neighbor Bruce, who recently came down with a rash, aching joints, and low fever, the pathogen might have been an insect-borne bacterium picked up from his trip to Mexico, or Lyme disease from his own backyard, or something else entirely. His doctor had no way to know and so, bumping up against a long weekend, prescribed two weeks of doxycycline just to cover whatever undiagnosed bug, if any, Bruce might be harboring. If the culprit had been a virus, the antibiotic would have been useless. But what if, instead, Bruce's doctor had access to a machine that could spit out the name of pathogen X in a few hours—or even a few minutes? It is a different way of approaching the problem, says O'Grady. "You sequence and find out what's out there." Both O'Grady and Bergman think that this kind of rapid read is the way forward. (And as of 2007, "what's out there" could be any of the thousands of human, plant, and animal bacterial pathogens that have been sequenced.[26]) The MinION device, small enough to fit in a pocket, plugs into a laptop and within hours sequences enough DNA to accurately identify bacteria, viruses, and other organisms with relatively small genomes.[27]

For patients like Bruce who are treated without definitive diagnosis, and for the health care workers who must make the call to treat or not and what with, this kind of technology can't come too soon. As with any new technology, there are some kinks. For O'Grady, one problem is *too* much DNA. Consider blood. In one milliliter (less than a quarter of a teaspoon), we have billions of red blood cells, and hundreds of millions of white bloods cells—the immune cells that protect us upon invasion. Our red cells lack DNA, but our white blood cells are packed with it, each cell containing our entire human genome. Yet even with a raging infection, O'Grady says a sample can have only one to ten bacteria, each with a much smaller genome of its own. "So you've got this ridiculous 10 billion : 1 ratio and you need to reduce it, hopefully to a 10 : 1 ratio, and then you can sequence. We can do that," says O'Grady. "We can remove 99.9995 percent human DNA, and we detected from real patients . . . the pathogen *and* the antibiotic resistance genes."

Urinary tract infections (UTIs) are notorious for resistance, which is why they are a good target for antibacterial developers and why colistin resistance is such a frightening prospect. What's more, it can be difficult to tease apart those suffering from infections from those who are simply colonized by bacteria without ill effect. Because urine carries fewer human cells, though, it is also easier to tease out bacterial DNA compared with blood. O'Grady says it takes about an hour to effectively get rid of about 95 percent of human DNA. And, unlike sepsis, urinary tract infections are practically teeming with bacteria. "We can take a urine sample . . . , get rid of human DNA, take bacterial DNA, and sequence—and you have the pathogen and all the resistance genes in a total of four hours." The ability to rapidly sequence any and all DNA is a game-changer.

These agent-agnostic methods, says Nicholas Bergman, offer "extraordinarily powerful approaches able to detect essentially anything and everything, including a pathogen the world has never seen before." This characteristic may prove invaluable for halting an emerging pathogen before it goes pandemic. As Ebola was raging in West Africa during the 2015 outbreak, Nicholas Loman and colleagues who study the genomes of human diseases were shocked by the lack of available genomic sequences for the pathogen, even though over 15,000 cases had been reported. In response, they put MinION to the test in the field. "It seemed obvious to us that Ebola sequencing could be done on nanopore [that is, using nanopore technology] near the portable diagnostic laboratories set up across West Africa," writes Loman in a blog about the experience. Within twenty-four hours of sampling, the group was able to detect mutations in the virus. This revealed another powerful advantage of rapid sequencers. Many viruses, influenza in particular, are notorious for their ability to mutate, evading treatment and prevention. Capturing those changes in real time will be invaluable. Loman writes, "We stand now at a confluence of new technologies and ideas that are changing the way we can do epidemiology and public health surveillance."[28]

Loman's student Joshua Quick, who traveled to Guinea to sequence Ebola, says that the method is democratizing sequencing.[29] It may also give humanity an edge against invisible threats from near and far. "I think everyone expects that applications like biosurveillance will shift to genomics very soon," says Bergman. "I would actually argue that the 'detect anything' advantage makes it worth shifting now."

Midcourse Corrections?

Twenty-first-century diagnostics can move medicine into an age of *directed treatment*: antibiotics better suited for the infection, such as bacteriocins and phage therapies that target specific strains of bacteria while leaving others unharmed. We can go after our natural enemies without hurting our allies. Yet even as diagnostic tests prove increasingly accurate, there are still barriers to overcome. Inertia is a major one, particularly here in the United States and other developed countries where hospitals and clinics have the luxury of a lab a few floors or a short drive away. "They may have already invested x dollars in devices, equipment, or lab space," says Seila Selimovic, a physicist, engineer, and innovator in diagnostics and smart technologies.[30] "What will it take to convince them to move to a different device? They might think the more precise answer they get in their existing lab is what they need. US labs and offices are not really resource-poor. They have equipment that they're fine with. What is the incentive to move to rapid diagnostics tests?" Selimovic contrasts the ready availability of diagnostics here in the United States with the lack of available tests in developing countries, where a village clinic may be a day or two away from the lab: "This is a place where rapid diagnostics are absolutely needed."

And, adds Selimovic with a hint of resignation in her voice, "There is not necessarily a mistrust, but maybe a hesitance to adopt new technology. It's another thing you have to learn how to use, another code for people who have to deal with billing. There is this feeling of 'let's keep doing what we are doing because it works well' until you hit a wall.

And I don't know if we've already hit a wall with antibiotic-resistant pathogens."

Bergman, too, says that institutional and technological inertia may slow adoption. "It'll take some significant retooling of the detection/diagnostic labs in terms of equipment, training, and staffing, and that's expensive. So the bottom line is that everyone can see where we want to go, but no one knows quite what the road looks like."

If we are serious about preserving antibiotics, working with our microbiomes, and preventing epidemic disease from racing around the world, then rapid and accurate identification of bacteria, viruses, protozoans, and fungal invaders is critical. We need to protect and preserve our beneficial microbiome while zeroing in on opportunistic bugs like *C. diff*, MRSA, gonorrhea, and TB. These twenty-first-century technologies can bring us into a new age of diagnostics, one where we know our enemy and can protect against it by enlisting our natural allies.

Epilogue

"I was a Mississippian. I'd grown up there. I'd go to bed—this was before AC. Everybody would shut up the windows and spray the house with DDT," recalls scientist Claude Boyd. You can hear the Deep South in his voice as he reminisces. "Back then, the towns, they had a mosquito fogger and I guess they had DDT in that and they just went down the streets every afternoon. That kind of bothered me. . . . They just sprayed it everywhere back then."

Boyd went on to a long, distinguished career in aquaculture and is now professor emeritus at Auburn University. As a master's student, he collected frogs, fish, and birds from the bayous and marshes bordering pesticide-soaked cotton fields. As he counted the dead, he was finding some of the first indications that chemicals influence evolution in wildlife. Many of those pesticides are now banned, but in the 1940s and '50s, it was common for growers to spray them more than a dozen times a season.

Boyd, like my parents, grew up in the heyday of "Better Living through Chemistry." The chemical industry produced the medicines that saved my father's life. Its pesticides prevented mosquito-borne diseases and allowed growers to harvest cotton, corn, and wheat that was under siege from weevils, black rot, and corn borers. Boyd, for one, appreciates the value of twentieth-century innovations, telling me, "Certain compounds need to be done away with, but I think they [chemicals] are absolutely necessary for various things for food production and human health." But, he says, "they need to be used responsibly." He sees that responsibility in the kind of improved diagnostics and targeted treatments highlighted throughout this book.

Restraint and precision weren't part of the plan half a century ago, when Boyd set out for the cotton fields as an inquisitive young student. Today we recognize that we cannot simply beat back pests and pathogens by blasting them with chemicals. We need an approach that is better informed by ecology, biology, chemistry, and genetics. The goal is to keep antibiotics and pesticides effective, while maintaining our natural allies—from the microbes in our guts to the beneficial microbes and insects on the farm. We are changing our strategy from all-out warfare to managing a new understanding of "wildlife" that includes bugs and germs.[1]

New scientific technologies have introduced us to once invisible worlds. We always knew that we lived amidst myriad microbes, but now genomics has unveiled the identity of the trillions of bacteria living within us, upon us, and in the roots, shoots, and soils surrounding plants. There's no question that some of these microbes are dangerous, but most are beneficial for plants, humans, and other living things. Advances in genetic engineering are also creating new vaccines and plants modified to resist disease, reducing the need for fungicides.

In the previous chapters, I have highlighted a handful of strategies for protection, prevention, and diagnostics. Each chapter could easily have become a book of its own, because there are hundreds of ongoing studies to identify new antimicrobials, better vaccines, and faster diagnosis.

My goal was not to pick the winners, but to explore a promising shift in pest and disease management. By the time you read this book, some of these treatments may be available to health care providers and farmers; others will not, at least not in their current form. Perhaps a chemical, even if naturally derived, is too toxic; or it's just too expensive to produce; or a laboratory lacks the funds necessary to navigate expensive testing and trials; or a vaccine is effective but not effective enough; or a new diagnostic is overly sensitive. Yet these research efforts, at the very least, will lay the groundwork for the next generation of solutions. In the last century, we got schooled in the dangers of trying to dominate nature. This century provides new lessons:

There will always be pests and pathogens. There will be disease-causing bacteria like staph, TB, Ebola, and gonorrhea. There will be mosquitoes and aphids and moths. Farmers will forever be fending off insects and weeds. This is why we need to rethink our strategy of hammering pest and pathogen into oblivion. It doesn't work. They will survive. And even if we were to successfully destroy a target pest, another would take its place.

We must respect ecology. For too long we have separated ourselves from nature. Both human health and agricultural systems are stronger when they exist as part of a healthy, functioning ecosystem. Nature does not need us, but we do need nature.

We are just a few macroorganisms in a microbe-filled world. We are just beginning to discover the details about which bacteria live where and how they interact with others, but at the very least we now understand that the microbiome is integral to both human and agricultural health. New tools like genomics and metagenomics are helping us learn more. And as we do, we can develop better ways to support these natural systems.

We do not live in a black-and-white world. It can be tempting to shun any farmer who uses pesticides or to disparage doctors who prescribe antibiotics as a preventative measure. But such reactions are oversimplified and polarizing, even potentially dangerous. Life is complex. We can act with better information rather than react out of fear.

Nature's own are powerful allies. We can harness antimicrobials from bacteria; use viruses to attack bacteria; reproduce pheromones; and otherwise enlist bugs and germs for our benefit. Some natural solutions may be intrinsically safer than synthetic chemicals, but *natural* does not necessarily equal *safe*. Nature produces some of the most powerful toxins known.

Specificity reduces collateral damage. Targeted antibiotics and pesticides fight specific pests and pathogens while preserving the microbiome or ecosystem. But *specific* can be interpreted as *narrow*, discouraging private research and development and widespread adoption. Precision shouldn't be the antithesis of profit.

Rapid diagnostics are essential. The quicker a problem is identified, the more effectively it can be treated. We need to know who and what we are up against, particularly as we move toward more-targeted treatments. Diagnostics in medicine and on the field are critical to better solutions, particularly when expert advice isn't in the next town over, or is too expensive. Machine learning and genomic sequencing may revolutionize diagnostics.

There is no better cure than prevention. GMO can provide resistance genes, boosting a plant's response, while vaccines can provoke our own immune response. Prevention also reduces dependence on pesticides and antibiotics, helping to avoid resistance.

The best solutions aren't useful if they aren't available. Several scientists I interviewed believe Big Ag and Big Pharma are the key to getting their technologies widely adopted. These are the corporations we love to hate, but they have the funds, the laboratories, and the market access. If held accountable, corporations can lead the way to a less toxic future. Meanwhile many scientists, wary of the "Monsanto effect," are committed to keeping their products available and affordable without the help of private entities. A more responsive regulatory system could help keep costs down and provide options for development other than through corporations. Either way, solutions only work when they are used.

There are no miracle cures. We are flooded with headlines that tout the health benefits of coffee, vitamin C, or sunlight, and then the headlines flip when new studies show the opposite. Any one of the technologies described here could become a blazing headline one day (Microbes save the Farm! Vaccines on Demand!), and a disappointment the next. *Headlines are not science.* Independent of the news cycle, scientists move ahead, teasing out what works, and why—as well as what doesn't, and why not. Most of our solutions will come from this kind of incremental science.

On hearing that I was wrapping this book up, a friend congratulated me, saying, "You must be happy." Yes, I am, though I admit that finishing the book also makes me anxious—because science is never "finished." What I edited yesterday may be out of date in a month, or six months, or a year. Science keeps moving—and so quickly these days. But the book is less about any specific solution and more about the nature of our solutions. While Claude Boyd saw the worst of our chemical excesses in the field, I grew up at the tail end of the *Time-Life*, gee-whiz technology years. As we edged closer to a new millennium, the wonder products of my youth began to fail. Worse, they created disasters. I was part of the post–*Silent Spring* generation who questioned the chemical sins of our fathers, who questioned technology itself. But I believe that technology is now providing us with a new gift. It is the gift not of a miracle product, but of a new understanding. Technology is teaching us how complex the world is, how our environment is paradoxically resilient and fragile at the same time, and how our impact upon it is everlasting. This is a lesson in treading more lightly and choosing our battles. If we learn it, we can work *with* nature to protect our food and our health.

Notes

Chapter 1

1. Karen Anderson, in discussion with the author on March 15, 2015. Note that all quotes attributed to Karen are drawn from this conversation and follow-up e-mails.
2. Centers for Disease Control and Prevention, "*Clostridium difficile* Infection," http://www.cdc.gov/HAI/organisms/cdiff/Cdiff_infect.html, last updated March 1, 2016.
3. For a review see: Chandrabali Ghose, "*Clostridium difficile* in the Twenty-First Century," *Emerging Microbes and Infections* 2 (2013): e62, doi:10.1038/emi.2013.62.
4. Ibid.
5. For a review, see: Jens Walter and Ruth Ley, "The Human Gut Microbiome: Ecology and Recent Evolutionary Changes," *Annual Review of Microbiology* 65 (2011): 411–29.
6. University of Gothenburg, "Surface Area of the Digestive Tract Much Smaller than Previously Thought," *Science Daily*, April 23, 2014, https://www.sciencedaily.com/releases/2014/04/140423111505.htm.
7. Ron Sender, Shai Fuchs, Ron Milo, "Revised Estimates for the Number of Human and Bacteria Cells in the Body," *PLOS Biology* 14, no. 8

(August 2016), http://journals.plos.org/plosbiology/article?id=10.1371/journal.pbio.1002533, accessed December 5, 2016.

8. For a review, see: Clyde Huchinson III, "DNA Sequencing: Bench to Bedside and Beyond," *Nucleic Acids Research* 35, no. 18 (September 2007): 6227–37, doi:10.1093/nar/gkm688.
9. For a review, see: H. S. Bilofsky et al., "The GenBank Genetic Sequence Databank," *Nucleic Acids Research* 14 (1986): 1–4.
10. Dennis A. Benson et al., "GenBank," *Nucleic Acids Research* 42 (Database issue, 2014): D32–37; for more information, see: National Center for Biotechnology Information, US National Library of Medicine, "Genome," http://www.ncbi.nlm.nih.gov/genome/browse/, accessed August 3, 2016.
11. Jack Gilbert (Department of Surgery, University of Chicago, and founder of the Earth Microbiome Project), in discussion with the author, August 25, 2015. Note that all quotes attributed to Gilbert are drawn from this conversation and follow-up e-mails.
12. Louis Pasteur, "The Germ Theory and Its Applications to Medicine and Surgery" (1878), translated version available at http://www.bartleby.com/38/7/7.html, accessed August 3, 2016.
13. Klaus Strebhardt and Axel Ullrich, "Paul Ehrlich's Magic Bullet Concept: 100 Years of Progress," *Nature Reviews* 8 (June 2008): 473–80.
14. Louis Fischer, "Syphilis in Children," *Journal of the American Medical Association* 56 (1911): 406.
15. Ibid., 407.
16. N. Svartz of the Swedish Royal Caroline Institute, "The Nobel Prize in Physiology or Medicine 1939—Award Ceremony Speech," *Nobelprize.org*, Nobel Media AB 2014, http://www.nobelprize.org/nobel_prizes/medicine/laureates/1939/press.html, accessed August 3, 2016.
17. Quoted in: Jack A. Gilbert and J. D. Neufeld, "Life in a World without Microbes," *PLOS Biology* 12 (2014): e1002020, doi:10.1371/journal.pbio.1002020.
18. For a discussion of life without microbes, see previously referenced article by Gilbert and Neufeld.
19. Martin Blaser, *Missing Microbes* (New York: Picador Press, 2014).
20. Ibid., 119.

21. Harvard Health Men's Watch, "*Clostridium difficile*: An Intestinal Infection on the Rise," Harvard Health Publications, June 2010, http://www.health.harvard.edu/staying-healthy/clostridium-difficile-an-intestinal-infection-on-the-rise.
22. John G. Bartlett, "Historical Perspectives on Studies of *C. difficile* and *C. difficile* Infections," *Clinical Infectious Diseases* 46 (2008): S11.
23. Daniel A. Voth and Jimmy D. Ballard, "*Clostridium difficile* Toxins: Mechanism of Action and Role in Disease," *Clinical Microbiology Reviews* 18 (April 2005): 247–63.
24. Fernanda C. Lessa et al., "Burden of *Clostridum difficile* Infection in the United States," *New England Journal of Medicine* 372 (February 2015): 825–34.
25. Centers for Disease Control and Prevention, "*Clostridium difficile* Infections," *Health Care Associated Infections*, http://www.cdc.gov/HAI/organisms/cdiff/Cdiff_infect.html, updated March 1, 2016.
26. Peter Andrey Smith, "The Tantalizing Links between Gut Microbes and the Brain," *Nature News*, Nature Publishing Group, October 14, 2015, http://www.nature.com/news/the-tantalizing-links-between-gut-microbes-and-the-brain-1.18557.
27. Rob Knight, "How Our Microbes Make Us Who We Are," TED talk, posted February 2015, https://www.ted.com/talks/rob_knight_how_our_microbes_make_us_who_we_are/transcript?language=en.
28. National Center for Complementary and Integrative Health, "Probiotics in Depth," https://nccih.nih.gov/health/probiotics/introduction.htm#hed1, last updated July 2015.
29. US Food and Drug Administration, "Guidance for Industry: Enforcement Policy Regarding Investigational New Drug Requirements for Use of Fecal Microbiota for Transplantation to Treat *Clostridium difficile* Infection Not Responsive to Standard Therapies," US Department of Health and Human Services, http://www.fda.gov/downloads/BiologicsBloodVaccines/GuidanceComplianceRegulatoryInformation/Guidances/Vaccines/UCM488223.pdf?source=govdelivery&utm_medium=email&utm_source=govdelivery, July 2013.
30. Faming Zhang et al., "Should We Standardize a 1,700-Year-Old Treatment?" *American Journal of Gastroenterology* 107 (November 2012): 1755, doi:10.1038/ajg.2012.251.

31. Lawrence J. Brandt et al., "Long-Term Follow-Up of Colonoscopic Fecal Microbiota Transplant for Recurrent *Clostridium difficile* Infection," *American Journal of Gastroenterology* 107 (July 2012): 1079–87, doi:10.1038/ajg.2012.60; Ciarán P. Kelly, "Fecal Microbiota Transplantation—An Old Therapy Comes of Age," *New England Journal of Medicine* 368 (2013): 474–75.
32. For more about Open Biome, see: http://www.openbiome.org/about, accessed August 3, 2016.
33. "Fecal Transplants Shown Effective—No Mention of Ecology or Evolution," *Evolution and Medicine Review*, January 18, 2013, http://evmedreview.com/?s=fecal, accessed August 3, 2016.
34. Joshua Lederberg, "Infectious History," *Science* 288 (2000): 287–93.

Chapter 2

1. Daniel Geisseler and William R. Horwath, "Strawberry Production in California," http://apps.cdfa.ca.gov/frep/docs/Strawberry_Production_CA.pdf, May 2014.
2. Pesticide Action Network, "What's on My Food: Strawberries," http://www.whatsonmyfood.org/food.jsp?food=ST, accessed August 3, 2016.
3. For more, see EWG's 2016 "Shopper's Guide to Pesticides in Produce," https://www.ewg.org/foodnews/dirty_dozen_list.php, accessed November 30, 2016.
4. Margaret Lloyd (Small Farms Advisor, University of California), in discussion with the author, August 28, 2015. Note that all quoted material attributed to Lloyd is drawn from this discussion and follow-up e-mails.
5. The image accompanies an article by Mike Amaranthus and Bruce Allyn, "Healthy Soil Microbes, Healthy People," *Atlantic Monthly*, June 11, 2013, http://www.theatlantic.com/health/archive/2013/06/healthy-soil-microbes-healthy-people/276710/, accessed August 3, 2016.
6. Rodrigo Mendes et al., "Deciphering the Rhizosphere Microbiome for Disease-Suppressive Bacteria," *Science* 233 (2013): 1097–100.
7. Jop de Vrieze, "The Littlest Farmhands," *Science* 349 (2015): 680–83.
8. Marie Chave, Marc Tchamitchian, and Harry Ozier-Lafontaine, "Agroecological Engineering to Biocontrol Soil Pests for Crop Health," *Sustainable Agricultural Reviews* 14 (2014): 269–97.

9. Ibid.; see also: Rodrigo Mendes, Paolina Garbeva, and Jos Raaijmakers, "The Rhizosphere Microbiome: Significance of Plant-Beneficial, Plant-Pathogenic, and Human-Pathogenic Microorganisms," *Federation of European Microbiological Societies, Microbiology Reviews* 37 (2013): 634–63.
10. Tom Curtis, "Microbial Ecologists: It's Time to Go Large," *Nature Reviews Microbiology* 4 (July 2006): 488.
11. US Department of Agriculture, "U.S. Strawberry Consumption Continues to Grow," https://www.ers.usda.gov/data-products/chart-gallery/gallery/chart-detail/?chartId=77884, accessed November 30, 2016.
12. US Department of Agriculture, "Economic Implications of the Methyl Bromine Phaseout," *Agricultural Information Bulletin* 756 (Washington DC: USDA Economic Research Service, February 2000).
13. US Environmental Protection Agency, "Methyl Bromide," https://www.epa.gov/ods-phaseout/methyl-bromide, updated March 2016.
14. Steven A. Fennimore et al., "Methyl Bromide Alternatives Evaluated for California Strawberry Nurseries," *California Agriculture* 62 (2008): 62–67; USDA, "Economic Implications."
15. Sir Albert Howard, "Farming and Gardening for Health or Disease," in *Articles of Sir Albert Howard for Organic Gardening Magazine, 1945–1947*, http://soilandhealth.org/wp-content/uploads/01aglibrary/010142howard.misc/010140.ogf.1945-47.htm, accessed August 4, 2016.
16. Sir Albert Howard, *The Soil and Health* (Lexington, KY: University of Kentucky Press, 2006), 22–23.
17. David Montgomery and Anne Bikle, *The Hidden Half of Nature: The Microbial Roots of Life and Health* (New York: W. W. Norton, 2015), e-book, 89.
18. Ibid., 89.
19. Angela Sessitsch and Birgit Mitter, "21st Century Agriculture: Integration of Plant Microbiomes for Improved Crop Production and Food Security," *Microbial Biotechnology* 18 (2015): 32–33; see also: Peter Andrey Smith, "Why Tiny Microbes Mean Big Things for Farming," *National Geographic* (blog), September 18, 2014, http://news.nationalgeographic.com/news/2014/09/140918-soil-bacteria-microbe-farming-technology-ngfood/, accessed August 4, 2016.

20. Jesse Ausubel, Iddo Wernick, and Paul Waggoner, "Peak Farmland and the Prospect for Land Sparing," *Population and Development Review* 38, Supplement (2012): 221–42.
21. Food and Agriculture Organization of the United Nations, "2015 International Year of Soils," http://www.fao.org/soils-2015/en/?utm_source=faohomepage&utm_medium=web&utm_campaign=featurebar, updated November 2015.
22. US Department of Agriculture, "World Agriculture Supply and Demand Estimates," WASDE-55, July 12, 2016, http://www.usda.gov/oce/commodity/wasde/latest.pdf, accessed August 4, 2016.
23. Youn-Sig Kwak and David Weller, "Take-all of Wheat and Natural Disease Suppression: A Review," *Journal of Plant Pathology* 29 (2013): 125–35.
24. M. Gerlagh, "Introduction of *Ophiobolus graminis* into New Polders and Its Decline" (thesis, Center for Agricultural Publishing and Documentation, Wageningen, Netherlands, 1968), ISBN 9022001776, http://edepot.wur.nl/191755, accessed August 4, 2016.
25. Surendra Dara, Karen Klonsky, and Richard De Moura, "2011 Sample Costs to Produce Strawberries, South Coast Region," University of California Cooperative Extension publication, 2011, http://coststudyfiles.ucdavis.edu/uploads/cs_public/87/d1/87d1dc5f-60ea-453c-a349-ef0a8ce3a851/strawberry_sc_smv2011.pdf.
26. Ryan Voiland (owner of Red Fire Farm, Montague, MA), in discussion with the author, November 5, 2014. Note that all quoted material attributed to Voiland is drawn from this discussion and follow-up e-mails.
27. Daniel Chellemi et al., "Development and Deployment of Systems-Based Approaches for the Management of Soilborne Plant Pathogens," *Phytopathology* 106 (2016): 216–25; Daniel Chellemi, J. W. Noling, and S. Sambhav, "Organic Amendments and Pathogen Control: Phytopathological and Agronomic Aspects," in M. L. Gullino, M. Pugliese, and J. Katan (eds.), *Proceedings of the 8th International Symposium on Chemical and Non-Chemical Soil and Substrate Disinfestation* (2014): 95–103; Daniel Chellemi, E. N. Rosskopf, and N. Kokalis-Burelle, "The Effect of Transitional Organic Production Practices on Soilborne Pests of Tomato in a Simulated Microplot Study," *Phytopathology* 103 (2013): 792–891.

28. Daniel Chellemi (plant pathologist and former applied-research manager, Driscoll Strawberry Associates, Watsonville, CA), in discussion with author, September 1, 2015. Note that all quoted material attributed to Chellemi is drawn from this discussion and follow-up e-mails.
29. Jae-Yul Cha et al., "Microbial and Biochemical Basis of a Fusarium Wilt–Suppressive Soil," *International Society for Microbial Ecology Journal* 10 (2016): 119–29.
30. Ibid., 128.
31. International Development Research Center, "Facts and Figures on Food and Biodiversity," https://www.idrc.ca/en/article/facts-figures-food-and-biodiversity, accessed August 4, 2016.
32. Arthur Grube et al., "Pesticides Industry Sales and Usage—2006 and 2007 Market Estimates," US Environmental Protection Agency, February 2011, https://www.epa.gov/sites/production/files/2015-10/documents/market_estimates2007.pdf.
33. US Department of Agriculture, "Fertilizer Use and Price," http://www.ers.usda.gov/data-products/fertilizer-use-and-price.aspx#26720, updated July 2013; Food and Agriculture Organization of the United Nations, "Fertilizer Use to Surpass 200 Million Tonnes in 2018," http://www.fao.org/news/story/en/item/277488/icode/, accessed August 4, 2016.

Chapter 3

1. The Bacteriophage Ecology Group, "Phage Companies," http://companies.phage.org/, updated April 14, 2015.
2. Wendell Stanley, "The Isolation and Properties of Crystalline Tobacco Mosaic Virus," Nobel Media AB 2014, http://www.nobelprize.org/nobel_prizes/chemistry/laureates/1946/stanley-lecture.html.
3. Ibid.; see also: Laurence Moran, "Nobel Laureate: Wendell Stanley," *Sandwalk* (blog), May 28, 2008, http://sandwalk.blogspot.com/2008/05/nobel-laureate-wendell-stanley.html.
4. For a good overview of viruses, see: Carl Zimmer, *A Planet of Viruses* (Chicago: University of Chicago Press, 2012).
5. Louis Villarreal, "Are Viruses Alive?" *Scientific American*, December 2004.

6. Shira Abeles and David Pride, "Molecular Bases and the Role of Viruses in the Human Microbiome," *Journal of Molecular Biology* 429 (2014): 3892–906, http://dx.doi.org/10.1016/j.jmb.2014.07.002.
7. As quoted in: Paul Thacker, "Drug Resistance Renews Interest in Phage Therapy," *Journal of the American Medical Association* 290 (December 2003): 3183; also, for a good overview of phage therapy, see: Anna Kuchment, *The Forgotten Cure* (New York: Springer Press, 2012), 8.
8. Koren Wetmore, "A Cure Exists for Antibiotic-Resistant Infections: So Why Are Thousands of Americans Still Dying?" *Prevention*, January 1, 2015, http://www.prevention.com/health/health-concerns/cure-antibi otic-resistance. Also, see the website for the Phage Therapy Center in Tbilisi, Republic of Georgia: http://www.phagetherapycenter.com/pii /PatientServlet?command=static_ourhistory&language=0, accessed August 4, 2016.
9. Ibid.
10. Kuchment, *The Forgotten Cure*, 124.
11. Lawrence Goodridge, "Bacteriophages for Managing Shigella in Various Clinical and Nonclinical Settings," *Bacteriophage* 3 (2013): e25098; Alexander Sulakvelidze, Zemphira Alavidze, and J. Glenn Morris Jr, "Bacteriophage Therapy," *Antimicrobial Agents and Chemotherapy* 45 (2001): 649–59.
12. For an interesting exploration of why the West did not pursue phage therapy, see: Emiliano Fruciano and Shawna Borne, "Phage Therapy as an Antimicrobial Agent: d'Herelle's Heretical Theories and Their Role in the Decline of Phage Prophylaxis in the West," *Canadian Journal of Infectious Diseases and Medical Microbiology* 18 (2007): 19–26.
13. Ibid.
14. Elizabeth Kutter, speaking at the Eighth International Conference on Biotherapy, October 11–14, 2010, Hilton Universal City, California; all quotes from Kutter are drawn from this video: https://www.youtube .com/watch?v=_Ju8RelMQWc, accessed August 4, 2016.
15. Steven Abedon et al., "Phage Treatment of Human Infections," *Bacteriophage* 1 (2011): 66–85.
16. Suzanna (a pseudonym), in e-mail conversation with author, November 29, 2016.

17. For the AmpliPhi company website, see: http://www.ampliphibio.com/product-pipeline, accessed August 4, 2016; and for Phagoburn, see: http://www.phagoburn.eu/, accessed August 4, 2016.
18. Kelly Sevick, "Beleaguered Phage Therapy Trial Presses On," *Science* 352 (2016): 1506.
19. Randall Kincaid (Senior Scientific Officer, National Institutes of Health, Bethesda, MD), in discussion with the author, November 3, 2015. Note that all quoted material attributed to Kincaid is drawn from this discussion and follow-up e-mails.
20. Marco Ventura et al., "The Impact of Bacteriophages on Probiotic Bacteria and Gut Microbiota Diversity," *Genes and Nutrition* 6 (2011): 205–7; and for more about phages in general, see: "2015 The Year of the Phage," http://2015phage.org/index.php.
21. US Department of Health and Human Services, Food and Drug Association, "Food Additives Permitted for Direct Addition to Food for Human Consumption; Bacteriophage Preparation," Docket No. 2002F—0316, http://www.fda.gov/OHRMS/DOCKETS/98fr/cf0559.pdf; one recent preparation approved by the US FDA is SalmoFresh, a mix of phages aimed at salmonella bacteria in poultry and other foods.
22. Another *lack* of incentive for profit-motivated companies to develop natural organisms or products is that naturally existing organisms cannot be patented. (Phages and other biologicals altered by engineering, however, *can* be patented. And there is increasing interest in developing engineered phages.) But the regulatory procedures need to change as well to accommodate a treatment that may be tailored towards each patient, just as therapies and regulations must change in response to (or to avoid) evolution of the target pathogen (some have suggested a model akin to flu vaccines that are altered annually).
23. Margaret Riley (Department of Biology, University of Massachusetts), in discussion with the author, November 3, 2015. Note that all quoted material attributed to Riley is drawn from this discussion and follow-up e-mails.
24. B. C. Kirkup and M. Riley, "Antibiotic-Mediated Antagonism Leads to Bacterial Game of Rock-Paper-Scissors In-Vivo," *Nature* 428 (2004): 412–14.

25. See: Beth Ann Crozier-Dodson, Mark Carter, and Zouxing Zheng, "Formulating Food Safety: An Overview of Antimicrobial Agents," *Food Safety Magazine*, December 2004/January 2005, http://www.foodsafetymagazine.com/magazine-archive1/december-2004january-2005/formulating-food-safety-an-overview-of-antimicrobial-ingredients/; see also: http://www.fda.gov/ucm/groups/fdagov-public/@fdagov-foods-gen/documents/document/ucm266587.pdf, accessed August 5, 2016.
26. Medine Gulluce, Mehmet Karadayi, and Ozlem Baris, "Bacteriocins: Promising Natural Antimicrobials," in *Microbial Pathogens and Strategies for Combatting Them: Science, Technology, and Education*, ed. A. Mendez-Vilas (Badajoz, Spain: Formatex Research Center, 2013).
27. "Mastitis in Cattle," *The Merck Veterinary Manuel*, http://www.merckvetmanual.com/mvm/reproductive_system/mastitis_in_large_animals/mastitis_in_cattle.html, last updated October 2014; USDA Animal and Plant Health Inspection Service, Veterinary Services Centers for Epidemiology and Animal Health, "Prevalence of Contagious Mastitis Pathogens on U.S. Dairy Operations, 2007," https://www.aphis.usda.gov/animal_health/nahms/dairy/downloads/dairy07/Dairy07_is_ContMastitis.pdf, accessed August 5, 2016.
28. For more about the use of antibiotics in food animals and agriculture in general, see: Timothy Landers et al., "A Review of Antibiotic Use in Food Animals: Perspective, Policy, and Potential," *Public Health Reports* 127 (January/February 2012): 4–22; US Food and Drug Administration, "2009 Summary Report on Antimicrobials Sold or Distributed in Food-Producing Animals," released September 2014.
29. Margaret A. Riley et al., "Resistance Is Futile: The Bacteriocin Model for Addressing the Antibiotic-Resistance Challenge," *Biochemical Society Transactions* 40 (2012): 1438–42.
30. Centers for Disease Control and Prevention, "Catheter-Associated Urinary Tract Infections (CAUTI)," http://www.cdc.gov/HAI/ca_uti/uti.html, last updated October 16, 2015.
31. Xiao-Qing Qiu et al., "An Engineered Multidomain Bactericidal Peptide as a Model for Targeted Antibiotics Against Specific Bacteria," *Nature Biotechnology* 21 (2003): 1480–85.
33. Read all about it here: Gong Yidong, Eliot Marshall, "Doubts over New

Antibiotic Lands Co-Authors in Court," *Science* 311, no. 5763 (February 2006): 937.

33. "University Clears Biophysicist of Misconduct," *Science* 312 (2006): 511.
34. US Department of Health and Human Services, "Antibiotic Resistance Threats in the United States," Centers for Disease Control and Prevention, April 23, 2013, http://www.cdc.gov/drugresistance/threat-report-2013/pdf/ar-threats-2013-508.pdf#page=22.
35. For more about the effects of changing the human microbiome through medicine, see: Martin Blaser, *Missing Microbes* (New York: Henry Holt and Company, 2014).

Chapter 4

1. National Pesticide Information Center, "*Bacillus thuringiensis*: General Fact Sheet," http://npic.orst.edu/factsheets/BTgen.pdf, last reviewed February 2015.
2. Brian McSpadden Gardener (microbial ecologist and cofounder of 3Bar Biologics, Columbus, OH), in discussion with the author, August 15, 2015. Note that all quoted material attributed to McSpadden Gardener comes from this discussion and follow-up e-mails.
3. Bruce Caldwell (cofounder, 3Bar Biologics, Columbus, OH), in discussion with the author, August 30, 2015. Note that all quoted material attributed to Caldwell is drawn from this discussion and follow-up e-mails.
4. Monsanto Company, "Microbials: Sustainable Solutions for Agriculture," http://www.monsanto.com/products/pages/microbials.aspx, accessed August 8, 2016.
5. R. Douglas Sammons, "BioDirect and Managing Herbicide-Resistant Amaranth sp.," presentation delivered at Resistance 2015 Meeting, Rothamsted Agricultural, Harpenden, Hertfordshire, England, September 14–16, 2015.
6. For more about olfaction and pheromones in insects, see: Lee Sela and Noam Sobel, "Human Olfaction: A Constant State of Change-Blindness," *Experimental Brain Research* 205 (2010): 13–29; "Pheromones in Insects," *Smithsonian*, Information Sheet Number 148, May 1999; Karl-Ernst Kaissling, "Pheromone Receptors in Insects," chap. 4 in

Neurobiology of Chemical Communication, ed. C. Mucignat-Caretta (Boca Raton, FL: CRC Press/Taylor & Francis, 2014).

7. Jean Henri Fabre, "The Great Peacock," chap. XI in *The Life of the Caterpillar*, trans. Teixiera de Mattos, (New York: Dodd, Mead and Company, 1916), 261–62, available online at http://www.eldritchpress.org/jhf/c11.html.
8. US Department of Agriculture, *Pesticide Data Program Annual Summary, Calendar Year 2013*, December 2014, http://www.ams.usda.gov/sites/default/files/media/2013%20PDP%20Anuual%20Summary.pdf.
9. Jon Clements (University of Massachusetts Extension, Amherst, MA), in discussion with the author November 19, 2015. Note that all quotes attributed to Clements are drawn from this interview and follow-up e-mails.
10. Jay Brunner et al., "Mating Disruption of Codling Moth: A Perspective from the Western United States," *International Organization for Biological Control West Palaearctic Regional Sectional Bulletin* 5 (2001): 207–25.
11. Alison Northcott, Canadian Broadcasting News, "How Pheromones Are Reducing Pesticide Use in Quebec Apple Orchards," http://www.cbc.ca/news/canada/montreal/pheremones-pesticides-apple-orchards-1.3585104, updated May 17, 2016.
12. Brad Higbee (Wonderful Orchards, Bakersfield, CA), in e-mail correspondence with the author, June 2016. (Note that subsequent quotes from Higbee are also drawn from this correspondence.)
13. Dennis Pollock, "Research Maximizing Navel Orangeworm Management in Pistachio, Almond," *Western Farm Press*, March 10, 2016, http://westernfarmpress.com/tree-nuts/research-maximizing-navel-orangeworm-management-pistachio-almond, accessed August 8, 2016; Brad Higbee, Charles Burks, and Joel Siegel, "Successful Control of Navel Orangeworm *Amyelois transitella* over Four Years Using Mating Disruption and Soft Insecticides in an Areawide Approach," conference presentation, Entomological Society of America Annual Meeting, Indianapolis, Indiana, December 13–16, 2009, https://www.researchgate.net/publication/267905043_Successful_control_of_navel_orangeworm_Amyelois_transitella_over_four_years_using_mating_disruption_and_soft_insecticides_in_an_areawide_approach, accessed August 8, 2016.

14. For a review, see: P. Witzgall, P. Kirsch, and A. Cork, "Sex Pheromones and Their Impact on Management," *Journal of Chemical Ecology* 36 (2010): 80–100.
15. Cam Oehlschlager (vice president, ChemTica Internacional, Costa Rica), in e-mail correspondence with author, November 2015.
16. John Pickett (Rothamsted Research, Harpenden, England), in discussion with author, September 15, 2015. Note that all quotes attributed to Pickett are drawn from this conversation and follow-up e-mails.
17. John Pickett, "Food Security: Intensification of Agriculture Is Essential, for Which Current Tools Must Be Defended and New Sustainable Technologies Invented," *Food and Energy Security* 2 (2013): 167–73.
18. "From Lab to Land: Women in 'Push-Pull' Agriculture," International Centre for Insect Physiology and Ecology, Nairobi, Kenya, 2015, ISBN 978–9966–063–08–3, http://www.push-pull.net/women_in_push-pull.pdf.
19. For more, see: "Push-Pull," Gatsby Charitable Foundation, http://www.gatsby.org.uk/africa/programmes/push-pull, accessed August 8, 2016; "The Push-Pull Farming System: Climate Smart Sustainable Agriculture for Africa," International Centre for Insect Physiology and Ecology, Nairobi, Kenya, 2015, ISBN 978–9966–063–06–9.
20. Jonathan A. Foley et al., "Solutions for a Cultivated Planet," *Nature* 478 (2011): 337–42.
21. For an interesting article on the topic, see: Nathaniel Johnson, "So Can We Really Feed the World? Yes, and Here's How," *Grist*, http://grist.org/food/so-can-we-really-feed-the-world-yes-and-heres-how/, February 10, 2015.

Chapter 5

1. Ryan Voiland (Red Fire Farm, Montague, Massachusetts), in discussion with the author, November 5, 2014. Note that all quoted material attributed to Voiland comes from this discussion and follow-up e-mails.
2. Martha Stewart, "The Tomato Blight in My Garden," *Martha Up Close and Personal* (blog), August 11, 2009, http://www.themarthablog.com/2009/08/the-tomato-blight-in-my-garden.html, accessed August 8, 2016.
3. Julia Moskin, "Outbreak of Fungus Threatens Tomato Crop," *New York*

Times, July 18, 2009, http://www.nytimes.com/2009/07/18/nyregion/18tomatoes.html.

4. Mathew Fischer et al., "Emerging Fungal Threats to Animal, Plant, and Ecosystem Health," *Nature* 484 (2012): 186–94.
5. William Fry (professor of plant pathology, School of Integrative Plant Science, Cornell University), in discussion with the author, January 15, 2015. Note that all quoted material attributed to Fry is drawn from this discussion and follow-up e-mails.
6. Jack Vossen (Senior Scientist, Department of Plant Breeding, Wageningen University and Research, Netherlands), in discussion with the author, January 22, 2016. Note that all quoted material attributed to Vossen is drawn from this discussion and follow-up e-mails.
7. Pests and pathogens continue to evolve; some crops like Bt require refuge areas—some crop or buffer which helps reduce evolution of resistant insects.
8. Union of Concerned Scientists, "Our History and Our Accomplishments," http://www.ucsusa.org/about/history-of-accomplishments.html#.V36NArgrKhc, accessed August 8, 2016.
9. Warren Leary, "Genetic Engineering of Crops Can Spread Allergies, Study Shows," *New York Times*, March 14, 1996, http://www.nytimes.com/1996/03/14/us/genetic-engineering-of-crops-can-spread-allergies-study-shows.html, accessed August 8, 2016.
10. Nobel Laureates Supporting Precision Agriculture (GMOs), "Support GMOs and Golden Rice," June 29, 2016, supportprecisionagriculture.org, accessed August 8, 2016.
11. For more about the conflict, see: Joel Achenbach, "107 Nobel Laureates Sign Letter Blasting Greenpeace Over GMOs," June 30, 2016, https://www.washingtonpost.com/news/speaking-of-science/wp/2016/06/29/more-than-100-nobel-laureates-take-on-greenpeace-over-gmo-stance/, accessed August 8, 2016; Greenpeace International, "All that Glitters Is Not Gold—The Truth about GE 'Golden Rice,'" http://www.greenpeace.org/international/en/campaigns/agriculture/problem/Greenpeace-and-Golden-Rice/, accessed August 8, 2016.
12. Natasha Gilbert, "Case Studies: A Hard Look at GM Crops," *Nature* 497 (2013): 24–26; Pamela Ronald and Raoul Adamchak, *Tomorrow's Table:*

Organic Farming, Genetics, and the Future of Food (New York: Oxford University Press, 2011); Joel Regenstein and Robert Blair, *Genetic Modification and Food Quality: A Down to Earth Analysis* (Hoboken, NJ: John Wiley & Sons, 2015).

13. For a discussion of mutation rates in humans, see: Laurence Moran, "Human Mutation Rates—What's the Right Number?" *Sandwalk* (blog), http://sandwalk.blogspot.com/2015/04/human-mutation-rates-whats-right-number.html, accessed August 8, 2016.
14. Paige Johnson, *Garden History Girl* (blog), "Atomic Gardens," http://gardenhistorygirl.blogspot.com/2010/12/atomic-gardens.html, posted December 2, 2010.
15. For more about mutation breeding, see: Committee on Identifying and Assessing Unintended Effects of Genetically Engineered Food on Human Health, "Unintended Effects from Breeding," chap. 3 in *Safety of Genetically Engineered Foods* (Washington, DC: National Academies Press, 2004), http://www.nap.edu/read/10977/chapter/5#45.
16. For example, a gene for antibiotic resistance may be used for GM as part of the process; antibiotic resistance enables developers to more easily identify genetically transformed crops.
17. US Department of Agriculture, "J. R. Simplot Potato Petition to Extend Determination," Docket no. APHIS-2015–0088, https://www.aphis.usda.gov/aphis/ourfocus/biotechnology/SA_Environmental_Documents/SA_Environmental_Assessments/Simplot-Potato-3, last modified January 13, 2016.
18. For a history of genetic engineering in agriculture and tomatoes in particular, see: Daniel Charles, *Lords of the Harvest: Biotech, Big Money, and the Future of Food* (New York: Perseus Books, 2001); Paul Lewis, letter to the editor, *New York Times*, June 16, 1992, http://www.nytimes.com/1992/06/16/opinion/l-mutant-foods-create-risks-we-can-t-yet-guess-since-mary-shelley-332792.html, accessed September 26, 2016.
19. Charles, *Lords of the Harvest.*
20. Ibid., 42.
21. Wilhelm Klumper and Martin Qiam, "A Meta-Analysis of the Impacts of Genetically Modified Crops," *PLOS One*, November 13, 2011, dx.doi.org/10.1371/journal.pone.0111629, accessed August 8, 2016; Charles

Benbrook, "Impacts of Genetically Engineered Crops on Pesticide Use in the U.S.—The First Sixteen Years," *Environmental Sciences Europe* 24 (2012): 24.

22. Regenstein and Blair, *Genetic Modification*, 2.
23. Ronald and Adamchak, *Tomorrow's Table*.
24. University of Florida, "UF Creates Trees with Enhanced Resistance to Greening," https://news.ifas.ufl.edu/2015/11/uf-creates-trees-with-enhanced-resistance-to-greening/, posted November 2015; Texas Citrus Greening, "List of Materials Available for Asian Citrus Psyllid Control in Various Ecological Settings in Texas," http://www.texascitrusgreening.org/psyllid-control-treatments/materials/, accessed August 8, 2016; for a more complete article on GMOs for citrus greening, see: Amy Harmon, "A Race to Save the Orange by Altering Its DNA," *New York Times*, July 27, 2013.
25. Daniel Cressey, "GM Wheat That Emits Pest Alarm Signals Fails in Field Trials," *Nature*, June 25, 2015, doi:10.1038/nature.2015.17854.
26. For more about techniques, see: "Q&A with Haven Baker on Simplot's Innate Potatoes," *Biology Fortified* (blog), https://www.biofortified.org/2013/05/qa-with-haven-baker-innate-potatoes/, accessed August 8, 2016; Dan Charles, "GMO Potatoes Have Arrived but Will Anyone Buy Them?" *All Things Considered*, National Public Radio, January 13, 2015, http://www.npr.org/sections/thesalt/2015/01/13/376184710/gmo-potatoes-have-arrived-but-will-anyone-buy-them, accessed August 8, 2016.
27. EFSA Panel on Genetically Modified Organisms, "Scientific Opinion on Addressing the Safety Assessments of Plants Developed Through Cisgenesis and Intragenesis," *EFSA Journal* 10 (2012): 2561, doi:10.2903/j.efsa.2012.2561.
28. Simplot, "Innate Second-Generation Potato Receives FDA Safety Clearance," http://www.simplot.com/news/innate_second_generation_potato_receives_fda_safety_clearance, accessed August 8, 2016.
29. John Travis, "Making the Cut," *Science* 350 (2015): 1456–57.

Chapter 6

1. John Brown and Philip Condit, "Meningococcal Infections—Fort Ord and California," *California Medicine* 102 (1965): 171–80; Andrew W. Artenstein, Jason M. Opal, Steven M. Opal, Edmund C. Tramont,

Georges Peter, and Phillip K. Russell, "History of U.S. Military Contribution to Vaccines Against Infectious Diseases," *Military Medicine* 170 (2005): 3–11.

2. Centers for Disease Control and Prevention, "Revised Recommendations of the Advisory Committee on Immunization Practices to Vaccinate All Persons 11–18 with Meningococcal Conjugate Vaccine," *MMWR Weekly* 56 (2007): 794–95.
3. Centers for Disease Control and Prevention, "Community Settings as Risk Factor," http://www.cdc.gov/meningococcal/about/risk-community.html, updated October 22, 2015.
4. Ryan Jaslow, "Meningitis Strain from Princeton University Outbreak Kills Drexel Student," *CBS News*, March 18, 2014, http://www.cbsnews.com/news/meningitis-strain-from-princeton-university-outbreak-kills-drexel-student/, accessed August 9, 2016.
5. Josh Logue, "Meningitis Risks," *Inside Higher Ed*, February 8, 2016, https://www.insidehighered.com/news/2016/02/08/meningitis-three-campuses-leads-one-outbreak-and-one-death, accessed August 9, 2016.
6. Michael Woroby, Adam Bjork, and Joel Wertheim, "Point, Counterpoint: The Evolution of Pathogenic Viruses and Their Human Hosts," *Annual Review of Ecology, Evolutionary and Systematics* 38 (2007): 515–40.
7. These are *estimates* based on available mortality data compiled by David McCandless for an infographic: "20th Century Deaths, Selected Major Causes," http://infobeautiful3.s3.amazonaws.com/2013/03/iib_death_wellcome_collection_fullsize.png, accessed August 9, 2016. The infographic data, which are drawn from the World Health Organization and other sources, can be accessed here: https://docs.google.com/spreadsheets/d/1vHTUaPWwlCPCg4O5YvDFRuDOCA8xX2ERnJQImnps6sk/edit#gid=21.
8. For an interesting visualization, see: Neil Halloran, "Fallen," http://www.fallen.io/ww2/, accessed August 9, 2016.
9. World Health Organization, "Immunization Coverage," http://www.who.int/mediacentre/factsheets/fs378/en/, updated July 2016.
10. *Los Angeles Times* (from Associated Press), "Last U.S. Smallpox Victim Leaves Mental Scars on Witnesses," December 26, 2001, http://articles.latimes.com/2001/dec/26/news/mn-18048, accessed August 9, 2016.

11. World Health Organization, "Frequently Asked Questions and Answers on Smallpox," http://www.who.int/csr/disease/smallpox/faq/en/, updated June 28, 2016.
12. Josh Earnest, White House Daily Press Briefing, February 8, 2016, https://www.whitehouse.gov/the-press-office/2016/02/09/daily-press-briefing-press-secretary-josh-earnest-282016, accessed August 9, 2016.
13. Centers for Disease Control and Prevention, "U.S. Vaccines," Appendix B-2, April 2015, http://www.cdc.gov/vaccines/pubs/pinkbook/downloads/appendices/B/us-vaccines.pdf, accessed August 9, 2016.
14. For a video story by Maurice Hilleman, see: The College of Physicians of Philadelphia, "Mumps: Jeryl Lynn Story," *The History of Vaccines*, October 29, 2004, http://www.historyofvaccines.org/content/mumps-jeryl-lynn-story, accessed August 9, 2016.
15. Wayne C. Koff et al., "Accelerating Next Generation Vaccine Development for Global Disease Prevention," *Science* 340 (2013), doi:10.1126/science.1232910, accessed August 9, 2016.
16. Vincent Rancaniello, *Virology* (blog), http://www.virology.ws/2010/04/13/poliovirus-vaccine-sv40-and-human-cancer/, accessed October 2016.
17. Polio Global Eradication Initiative, "Vaccine-Derived Polio Viruses," http://www.polioeradication.org/polioandprevention/thevirus/vaccinederivedpolioviruses.aspx, accessed August 9, 2016.
18. Timothy Guzman, "Big Pharma, Big Profits: The Multibillion-Dollar Vaccine Market," *Global Research*, January 26, 2016, http://www.globalresearch.ca/big-pharma-and-big-profits-the-multibillion-dollar-vaccine-market/5503945Q2, accessed August 9, 2016.
19. Leonard Moise (EpiVax, Providence, RI), in discussion with the author, March 14, 2015. Subsequent quotations are drawn from this discussion and follow-up e-mails.
20. The vaccines we receive against diphtheria and tetanus also contain a single antigen active against the toxin they produce. If we become infected with tetanus or diphtheria, the bacteria themselves aren't likely to kill us, but the toxins they release can. Antibodies neutralize the toxins. But hepatitis B vaccine is something entirely different.
21. Centers for Disease Control and Prevention, "Hepatitis B," chap. 10 in *Epidemiology and Prevention of Vaccine-Preventable Diseases*, 13th ed., ed.

J. Hamborsky, A. Kroger, and S. Wolfe (Washington, DC: Public Health Foundation, 2015), http://www.cdc.gov/vaccines/pubs/pinkbook/downloads/hepb.pdf, accessed August 9, 2016.

22. Margie Patlak, with the assistance of Drs. Baruch Blumberg, Maurice Hilleman, and William Rutter, "The Hepatitis B Story," for *Beyond Discovery: The Path from Research to Human Benefit*, National Academy of Sciences, February 2000, http://www.nasonline.org/publications/beyond-discovery/hepatitis-b-story.pdf, accessed August 9, 2016.
23. Philip Boffey, "U.S. Approves a Genetically Altered Vaccine," *New York Times*, June 24, 1986, http://www.nytimes.com/1986/07/24/us/us-approves-a-genetically-altered-vaccine.html, accessed December 4, 2016.
24. Taylor & Francis Online, "'Immune Camouflage' May Explain H7N9 Influenza Vaccine Failure" *ScienceDaily*, September 24, 2015, https://www.sciencedaily.com/releases/2015/09/150924112532.htm, accessed August 9, 2016.
25. For a review of Moise's work, see: Leonard Moise et al., "Harnessing the Power of Immunoinformatics to Produce Improved Vaccines," *Expert Opinion in Drug Discovery* 6 (2011): 9–15, doi:10.1517/17460441.2011.534454.2011.
26. M. Pizza et al., "Identification of Vaccine Candidates Against Serogroup Meningococcus B by Whole Genome Sequencing," *Science* 287 (2000): 1816–20. Pizza not only pioneered the Men B vaccine, but also used genetic engineering to develop a safer whooping cough vaccine; for more about Pizza, see: "Mariagrazia Pizza," American Society for Microbiology, http://academy.asm.org/index.php/fellows-info/aam-fellows-elected-in-2015/5387-mariagrazia-pizza, accessed August 9, 2016.

Chapter 7

1. Reviewed in: E. C. Oerke, "Crop Losses to Pests," *Journal of Agricultural Sciences* 144 (2006): 31–43.
2. Gustavo Ferreira and Agnes Perez, US Department of Agriculture, "Fruit and Tree Nut Outlook, U.S. Citrus Crop Continues to Decline, 2015/16," FTS 361, March 31, 2016, https://www.ers.usda.gov/webdocs/publications/fts361/57076_fts-361-revised.pdf, accessed November 30, 2016.

3. For a review of extension programs in the United States, see: Sun Ling Wang, "Cooperative Extension System: Trends and Economic Impacts on U.S. Agriculture," *Choices* (2014): 1, http://choicesmagazine.org/choices-magazine/submitted-articles/cooperative-extension-system-trends-and-economic-impacts-on-us-agriculture; Alan Jones, "Opinion: The Planet Needs More Plant Scientists," *The Scientist*, October 2014, http://www.the-scientist.com/?articles.view/articleNo/41133/title/Opinion--The-Planet-Needs-More-Plant-Scientists/, accessed August 9, 2016.
4. A number of books provide a history of blight; one is Susan Bartoletti's *Black Potatoes: The Story of the Great Irish Potato Famine* (Boston: Houghton Mifflin Harcourt, 2005).
5. Charles Mann, "How the Potato Changed the World," *Smithsonian Magazine*, November 2011, http://www.smithsonianmag.com/history/how-the-potato-changed-the-world-108470605/?no-ist=&no-cache=&page=3, accessed August 9, 2016.
6. "Views of the Famine," *Illustrated London News*, February 13, 1847, https://viewsofthefamine.wordpress.com/illustrated-london-news/sketches-in-the-west-of-ireland/, accessed November 30, 2016.
7. Adapted from an article by E. G. Ruestow, "Anton von Leeuwenhoek and His Perception of Spermatozoa," *Journal of the History of Biology* 16 (1983): 185–224, http://10e.devbio.com/article.php?id=65, accessed August 9, 2016.
8. U. Kutschera and U. Hossfeld, "Physiological Phytopathology: Origin and Evolution of a Scientific Discipline," *Journal of Applied Botany and Food Quality* 85 (2012): 1–5, http://www.biodidaktik.uni-jena.de/imndipmedia/Publikationen+UH/Art01_Kutschera_Farbe.pdf, accessed August 9, 2016.
9. Ibid., 3.
10. Amir Nezhad, "Future of Portable Devices for Plant Pathogen Diagnosis," *Lab on a Chip* 14 (2014): 2887–904.
11. David Hughes (assistant professor of entomology and biology, Pennsylvania State University), in discussion with the author, April 5, 2015. Note that all quotes attributed to Hughes are drawn from this interview and follow-up e-mails.
12. British Society for Plant Pathology, "Plant Pathology and Education in

the UK: An Audit," 2012, 6, http://www.bspp.org.uk/society/docs/bspp-plant-pathology-audit-2012.pdf, accessed August 9, 2016.

13. Jones, "Opinion."
14. National Science Foundation, "Science and Engineering Indicators, 2014," chap. 2 in *Graduate Education, Enrollment and Degrees in the United States*, http://www.nsf.gov/statistics/seind14/index.cfm/chapter-2/c2s3.htm#s3, accessed August 9, 2016; Jones, "Opinion."
15. Thomas Gordon (professor of plant pathology, College of Agricultural and Environmental Sciences, University of California–Davis), e-mail correspondence with author, June and July 2016. All quotes are drawn from these e-mails.
16. Pierre Lebarth and Catherine Laurent, "Privatization of Agricultural Extension Services in the EU: Towards a Lack of Adequate Knowledge for Small-Scale Farms," *Food Policy* 38 (2013): 240–52; O. J. Saliu and A. Age, "Privatization of Agricultural Extension Services in Nigeria, Proposed Guidelines for Implementation," *American-Eurasian Journal of Sustainable Agriculture* 3 (2009): 332.
17. Marcel Salathé, "Open Data: Our Best Guarantee for a Just Algorithmic Future," *Marcel Salathé's Blog*, February 10, 2016, http://blog.salathe.com/open-data-our-best-guarantee-for-a-just-algorithmic-future, accessed August 9, 2016.
18. Gwyn E. Jones, "The Clarendon Letters," *Progress in Rural Extension and Community Development*, vol. 1 (New York: John Wiley and Sons, 1982), 16–17, file:///C:/Users/Emily/Downloads/Clarendon%20Letter%20.pdf, accessed August 9, 2016.
19. Gwyn Jones and Chris Garforth, "The History, Development, and Future of Agricultural Extension," chap. 1 in *Improving Agricultural Extension*, ed. Curtis Swanson, Robert Bentz, and Andrew Sofranko (Rome: UN Food and Agriculture Organization, 1997), http://www.fao.org/docrep/w5830e/w5830e03.htm, accessed August 9, 2016.
20. Margaret Lloyd (Small Farms Advisor, University of California, Davis), e-mail exchange with author, May/June 2016. Note that all quotes attributed to Lloyd are drawn from these e-mails. See also: John Tibbitts, "Extension, Beyond Traditional Academic Jobs," *Science*, May 16, 2016, http://www.sciencemag.org/careers/2016/05/extension-beyond-tradi

tional-academic-jobs, doi:10.1126/science.caredit.a1600078, accessed August 10, 2016.

21. As quoted in a news release from the École Polytechnique Fédérale de Lausanne, "Smartphones to Battle Crop Diseases," November 11, 2015, http://actu.epfl.ch/news/smartphones-to-battle-crop-disease/.
22. Pew Research Center, "Cell Phones in Africa: Communication Lifeline," April 2015, http://www.pewglobal.org/2015/04/15/cell-phones-in-africa-communication-lifeline/; Pew Research Center, "Smartphone Ownership and Internet Usage Continues to Climb in Emerging Economies," February, 2016, http://www.pewglobal.org/2016/02/22/smartphone-ownership-and-internet-usage-continues-to-climb-in-emerging-economies/.
23. "Ericsson Mobility Report on the Pulse of the Networked Society," February 2016, http://www.ericsson.com/res/docs/2016/mobility-report/ericsson-mobility-report-feb-2016-interim.pdf.
24. Lloyd.
25. Alan Turing, "Computing Machinery and Intelligence," *Mind* 59 (1950): 460.
26. Neeraj Kumar et al., "Leafsnap: A Computer Vision System for Automatic Plant Species Identification," in "Computer Vision—ECCV 2012 Series Lecture Notes," *Computer Science* 7573 (2012): 502–16.
27. Leafsnap dataset, http://leafsnap.com/dataset/, accessed August 10, 2016.
28. Marcel Salathé (associate professor, School of Life Sciences, School of Computer and Communication Sciences, École Polytechnique Fédérale Lausanne) in correspondence with the author. Note that all quotes attributed to Salathé are drawn from this correspondence, unless otherwise noted.

Chapter 8

1. E. Sapi et al., "Improved Culture Conditions for the Growth and Detection of Borrelia from Human Serum," *International Journal of Medical Sciences* 10 (2013): 362–76, doi:10.7150/ijms.5698, available from http://www.medsci.org/v10p0362.htm.
2. Patrick McGann et al., "*Escherichia coli* Harboring mcr-1 and blaCTX-M on a Novel IncF Plasmid: First Report of mcr-1 in the USA," *Antimicrobial Agents and Chemotherapy*, AAC-Accepted Manuscript Posted

Online 26 May 2016, http://aac.asm.org/content/early/2016/05/25/AAC.01103-16.full.pdf+htmldoi:10.1128/AAC.01103-16.

3. "Longevity Bulletin," from the Institute and Faculty of Actuaries, *Antimicrobial Resistance* 8 (May 2016), ISSN 2397–7221, Longevity%20Bulletin%20Issue%208%20FINAL%20FOR%20ONLINE.pdf, accessed August 10, 2016; Centers for Disease Control and Prevention, "Antibiotic and Antimicrobial Resistance," http://www.cdc.gov/drugresistance/, updated July 14, 2016.
4. National Institute of Allergy and Infectious Diseases, "Antimicrobial (Drug) Resistance—Quick Facts," https://www.niaid.nih.gov/topics/antimicrobialresistance/understanding/Pages/quickFacts.aspx, updated March 20, 2014.
5. Martin Blaser, "Antibiotic Use and Its Consequences for the Normal Microbiome," *Science* 352 (2016): 545.
6. Pranita D. Tamma and Sara E. Cosgrove, "Editorial: Addressing the Appropriateness of Outpatient Antibiotic Prescribing in the United States," *Journal of the American Medical Association* 315 (2016): 1839–41, doi:10.1001/jama.2016.428. (The authors of this editorial comment concluded that "these data are vulnerable to some important limitations largely related to necessary assumptions made by the investigators that may have led to an underestimation of the burden of inappropriate antibiotic use.")
7. For a review, see: Angela M. Caliendo et al., "Better Tests, Better Care: Improved Diagnostics for Infectious Diseases," *Clinical Infectious Diseases* 57, suppl. 3 (2013): S139–70.
8. Kalorama Information, "The World Market for Infectious Disease Diagnostic Tests," November 1, 2015, Pub ID: KLI5721838, http://www.kaloramainformation.com/Infectious-Disease-Diagnostic-9367616/, accessed August 10, 2016; for a report on why we need diagnostics, see: "Rapid Diagnostics Stopping Unnecessary Use of Antibiotics," *Review on Antimicrobial Resistance*, October 2015, http://amr-review.org/sites/default/files/Rapid%20Diagnostics%20-%20Stopping%20Unnecessary%20use%20of%20Antibiotics.pdf, accessed August 10, 2016.
9. Joseph Schwartzman (Dartmouth Hitchcock Medical Center, Lebanon, NH), in e-mail correspondence with the author, June 2016.
10. International Trade Administration, National Travel and Tourism Office,

"Fast Facts: United States Travel and Tourism Industry—2014," http://tinet.ita.doc.gov/outreachpages/download_data_table/Fast_Facts_2014.pdf, accessed August 10, 2016.

11. IPK International, "ITB World Travel Trends Report—2015/2016," http://www.itb-berlin.de/media/itbk/itbk_dl_all/itbk_dl_all_itbkongress/itbk_dl_all_itbkongress_itbkongress365/itbk_dl_all_itbkongress_itbkongress365_itblibrary/itbk_dl_all_itbkongress_itbkongress365_itblibrary_studien/ITB_World_Travel_Trends_Report_2015_2016.pdf, accessed November 10, 2016.
12. Kathleen McGraw (Chief Medical Officer, Brattleboro Memorial Hospital, Brattleboro, VT), in discussion with the author, April 16, 2016. Note that all quotes attributed to McGraw are drawn from this discussion and follow-up e-mails.
13. Christopher Appleton (Medical Director of Pathology, Brattleboro Memorial Hospital, Brattleboro, VT), in discussion with the author, May 17, 2016. Note that all quotes attributed to Appleton are drawn from this discussion and follow-up e-mails.
14. American Hospital Association, "Fast Facts on Hospitals," http://www.aha.org/research/rc/stat-studies/fast-facts.shtml, accessed August 10, 2016.
15. See: Eva Engvall, "Perspective on the Historical Note on EIA/ELISA by Dr. R.M. Lequin," *Clinical Chemistry* 51 (2005): 2225; see also: Bauke van Weeman, "The Rise of ELISA," *Clinical Chemistry* 51 (2005): 2226.
16. Van Weeman, "The Rise of ELISA."
17. Centers for Disease Control and Prevention, "Outbreaks of Respiratory Illness Mistakenly Attributed to Pertussis—New Hampshire, Massachusetts, and Tennessee, 2004–2006," *MMWR Weekly* 56 (2007): 837–42.
18. Centers for Disease Control and Prevention, "Pertussis (Whooping Cough): Best Practices for Healthcare Professionals on the Use of Polymerase Chain Reaction (PCR) for Diagnosing," http://www.cdc.gov/pertussis/clinical/diagnostic-testing/diagnosis-pcr-bestpractices.html, updated September 8, 2015.
19. Cepheid, "Our Mission," http://www.cepheid.com/us/about-us/inside-cepheid/our-misson, accessed August 10, 2016.
20. World Health Organization, "Largest Ever Roll-out of GeneXpert Rapid

TB Test Machines, in 21 Countries," http://www.who.int/tb/features_archive/xpertprojectlaunch/en/; for a study of the test's efficacy, see: U. B. Singh et al., "Genotypic, Phenotypic, and Clinical Validation of GeneXpert in Extra-Pulmonary and Pulmonary Tuberculosis in India," *PLoS ONE* 11 (2016): e0149258, doi:10.1371/journal.pone.0149258.

21. Justin O'Grady (Senior Lecturer in Medical Microbiology, Norwich Medical School, Norwich, UK), in discussion with the author, April 11, 2016. Note that all quotes attributed to O'Grady are drawn from this discussion and follow-up e-mails.
22. Caliendo et al., "Better Tests."
23. See the Biomerieux website: http://www.biomerieux-usa.com/clinical/vitek-2-healthcare, accessed August 10, 2016.
24. Making their system available to researchers to use, as MinION has done, contrasts with the more secretive start-up Theranos, a company that imploded under the weight of "disruptive" diagnostic promises that could not be fulfilled—including a laundry list of inexpensive diagnostics requiring only the smallest drop of blood, with some results available within hours. (There are many articles about the company and its downfall; see: Nick Stockton, "Everything You Need to Know about Theranos So Far," *Wired*, http://www.wired.com/2016/05/everything-need-know-theranos-saga-far/, accessed August 10, 2016.)
25. Nicholas Bergman (head of the Genomics Department at the Department of Homeland Security's National Biodefense Analysis and Countermeasures Center, Frederick, MD), in e-mail correspondence with the author, April 28, 2016. Note that all quotes attributed to Bergman are drawn from initial and follow-up e-mails.
26. Mark Pallen and Brenden Wren, "Bacterial Pathogenomics," *Nature* 449 (2007): 835–42, doi:10.1038/nature06248.
27. Erika Hayden, "Pint-Sized DNA Sequencer Impresses First Users," *Nature* 521 (2015): 15–16, doi:10.1038/521015a.
28. Nick Loman, "Behind the Paper: Real-Time Portable Sequencing for Ebola Surveillance," *Lab Blog*, February 3, 2016, http://lab.loman.net/2016/02/03/behind-the-paper-real-time-portable-sequencing-for-ebola-surveillance/, accessed August 10, 2016; see also: Lisa O'Carroll, "From Ebola to Zika, Tiny Mobile Lab Gives Real-Time DNA Data on

Outbreaks," *The Guardian*, February 3, 2016, https://www.theguardian.com/science/2016/feb/03/from-ebola-to-zika-tiny-mobile-lab-gives-real-time-dna-data-on-outbreaks; for the paper, see: Joshua Quick et al., "Real-time, Portable Genome Sequencing for Ebola Surveillance," *Nature* 530 (2016): 228–32.

29. Hayden, "Pint-Sized DNA Sequencer," 15.
30. Seila Selimovic (Program Director, Division of Discovery Science and Technology, National Institute of Biomedical Imaging and Bioengineering) in discussion with the author, April 2016. Subsequent quotations are drawn from this discussion.

Epilogue

1. Carl Zimmer, "Tending the Body's Microbial Garden," *New York Times*, June 18, 2012, http://www.nytimes.com/2012/06/19/science/studies-of-human-microbiome-yield-new-insights.html, accessed November 30, 2016.

Index

Island Press | Board of Directors